Everyday Discoveries

Everyday Discoveries

Amazingly Easy Science and Math Using Stuff You Already Have

Sharon MacDonald

Illustrations by Joan Waites

gryphon house

Beltsville, Maryland

To Louise Scanlon
who has been a guide, support and cheerleader.
Thanks!

Copyright © 1998 Sharon MacDonald
Published by Gryphon House, Inc.
10726 Tucker Street, Beltsville MD 20705

World Wide Web: http://www.ghbooks.com

Library of Congress Cataloging-in-Publication Data

MacDonald, Sharon.
    Everyday discoveries : amazingly easy science and math using stuff you already have / Sharon MacDonald.
      p. cm.
    Includes index.
    ISBN: 0-87659-196-9
    1. Science--Experiments. 2. Science--Study and teaching--Activity programs. 3. Mathematics--Experiments. 4. Mathematics--Study and teaching--Activity programs. I. Title
Q164.M22 1998
507.8--dc21   98-17151
                                    CIP

Table of contents

Preface

I want to share with you an experience I had that led me to write this book.

My husband George and I had a beautiful backyard planted with trees, shrubs, flowers and grass from the back door to the fence beyond. Many of the plants we had picked ourselves and watched grow over the years. We enjoyed feeding the birds and other wild animals that visited our lovely yard.

Our backyard presented a problem, however. We could not see it from the inside of our house. So we decided to add a sunroom with a row of windows to give us a better view. We met with the architect, designed our sunroom and waited for construction to begin.

The contractor showed up with his crew and took down our six-foot-high cedar fence. Many of our plants were along this fence, but those that survived were soon flattened by the dump truck as it drove into our yard. The men unloaded wood scraps, a rusted roll of wire and other strange tools. They worked for a full day scraping the grass from the construction site and digging a big wide hole. I did not recognize any of the tools except the hammers.

Then it rained for several days. Gone was our beautiful yard with its green grass and mature plants; in its place was just an ugly mud pit. Back came the work crew with the sun. The men whistled, shouted and banged amidst the mud and machinery noise. They were having a great time, but what I saw was not at all what I had imagined "the beginning" would look like.

The next day came the cement truck. It couldn't be driven into the yard, so the crew set up several open troughs for the cement to run through. Suddenly our backyard looked like some kind of ore mining operation. As the trenches filled with concrete, the mud pit disappeared under a carpet of wet concrete and stony drippings fell outside the trough. The workmen jumped in the concrete with high-top rubber boots, and they danced around in the cement with odd-looking, rake-like tools called "jitterbugs." They yelled a lot to the cement truck driver. The concrete came faster, then slower, then not at all. As I watched the concrete dry fast and hard and flat, I remember thinking that I hoped they knew what they were doing.

When the crew finally left and we dared to look outside, we saw a shinny, concrete slab with all the wooden stakes sticking up around it. Could this really be the beginning of our sunroom? George stood beside me and said, "Sharon, this is just the foundation. It is the most important part of our room. We want a good foundation—one that will last a lifetime."

What an eye opener. While our sunroom would one day be a beautiful room, the foundation was ugly. The process was loud, messy and disorganized. At that point I was struck with a thought that eventually became the reason I wrote this book: I was a foundation worker! My teaching was about building foundations. Not the kind sunrooms sit on, but the kind children stand on as their education is erected.

The foundation part is the messy, noisy, sometimes confusing phase of young children's growth and development. The tools, methods and materials are so different from those we see in later stages of development that we may not recognize them as the beginning for what comes later.

As foundation workers, we are building a foundation that will last children a lifetime.

Chapter 1

Science and Math: Building a Foundation

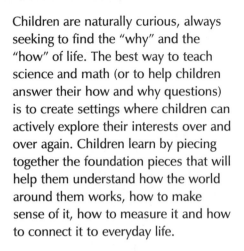

Children are naturally curious, always seeking to find the "why" and the "how" of life. The best way to teach science and math (or to help children answer their how and why questions) is to create settings where children can actively explore their interests over and over again. Children learn by piecing together the foundation pieces that will help them understand how the world around them works, how to make sense of it, how to measure it and how to connect it to everyday life.

You cannot have science without math because all science needs to be measured in one form or another. Sometimes the math foundations are obvious, sometimes they are not, but in each science activity there is math. Children can begin to connect science and math in the prekindergarten years. This is not difficult because science and math are present in every center and interest area, embedded in the activities. The pages of this book are filled with examples of how science and math are connected, but let's look at one example.

A simple problem: How many pumpkin seeds do we need for each child in the class to plant two seeds? The children figure out a way to count (one-to-one correspondence, grouping, etc.) and come up with an answer. They are learning or practicing math skills. Why? Because the children **want** to plant pumpkins. The mathematics skill of counting was necessary (and useful!) to plant pumpkin seeds—something known, something interesting (to them) and something concrete (part of their present experience). Additional math skills needed for the activity include figuring out **how deep** to dig the hole for the pumpkin seeds? **how far** apart to plant them? **how much** water to make them grow? As the first sprout appears: **how much** sunlight is needed? **how often** to water? **how much** it is growing? This easy and fun activity begins to build the foundations and combinations for children to acquire science and math within a frame of reference that interests them and that is part of their experience. Sometimes developing children's science and math skills is just a matter of recognizing the science and math in everything you do all day.

Everyday Discoveries shows you how to teach science and math easily in your classroom, following the guidelines of The National Council for Teachers of Mathematics (*Standards*) and the American Association for the Advancement of Science (*Benchmarks*). *Everyday Discoveries* highlights how easily, and naturally, children can learn science and math in early primary classrooms, child care centers, nursery schools and home care settings.

Foundation Tools

Teachers of young children use tools for learning that may not be seen in later years of education. Some examples are painting, drawing, singing, cooking, constructing, woodworking, pounding, kneading, pouring, dancing, jumping, reading, imagining, planning, thinking and dramatizing. Let's look at painting and see how it is a foundation tool for learning science and math. When children paint, many things are happening at once. They are exploring different amounts of paint on the brush; the size of the brush; gravity acting on the applied paint and the drips of paint, pulling them downward; and the friction created when they move a paint-filled brush across paper and other surfaces. As they paint over and over, they realize that they can control the volume of paint on the brush by putting less paint on the brush. They can select a fat or a thin brush to change the look of the paint on the paper. They can use the brush to stop the paint from dripping down the paper, and they can move the brush purposefully to create specific effects. Lots of science and math learning opportunities! Painting lets children explore volume, size, gravity and cause and effect. Volume and size are math skills. The science of gravity is a more difficult concept to grasp, but the experience of painting is a good foundation for understanding it later. Cause and effect shows children: if I do this (cause), that (effect) will happen.

Through repetition and by exploring things that are meaningful to them, children begin to use math to explore basic science concepts. They will build upon these later. The example above shows the potential for teaching these subjects. Science and math are everywhere, all you need to do is look for them.

Foundation Methods

Science and Math in All Centers

The methods used in early childhood settings are different than those seen in later years. Children are immersed in science and math when they are in the Art Center, the Science and Math Center, the Library Center, the Blocks Center, the Woodworking Center, the Sand and Water Center and Home-living Center. Science and math concepts and skills are interwoven throughout these settings. Let's look at one area, the Home-living Center, and see how naturally children learn the necessary skills and concepts through their play. This center invites children to come and play. They set the table, cook, dress-up, and sweep or mop. When the children set the table they are learning one-to-one correspondence. Obviously, this is math, but so is the **pattern** of each place setting, and **similarities** and **differences** between objects in the **grouping**. We know that young children cannot master math if they cannot see groups and the similarities and differences in groups (like groups of numbers).

But there's lots of science in the Home-living Center, too. When the children cook in the Home-living Center, they use measuring tools and they use artificial commodities—substitutes for the "real thing"—to represent food. This is science. Accepting **representations** of things as being the "real thing" is fundamental to scientific thinking and the methods of science. The artificial onion represents the real one. (Of course this is also important in math, for what is a number but a representation of something. Children must accept the convention that a number represents an object. Wouldn't schools be very large and noisy places if we had to have elephants in classrooms in order to count them!) When they sweep or mop, pattern, repetition and **force** (of the mop against the floor, for example) are explored. **Simple machines** are used, like the wedge and the lever. This is science (and math, too).

If teachers focus on science and math in a variety of centers, children will be immersed in science and math every day in a way that is meaningful for them. Their interests tell you what has meaning, and using what is meaningful for them is the "delivery system" for acquiring science and math skills.

Individual Learning Style

Children have different ability levels, different interests and different backgrounds that impact how they learn. Some children develop skills rapidly and others more slowly. Each child learns a little differently from another, too. Centers need to accommodate each child's individual learning style. How do you do it? Through choice. The children need lots and lots of activity choices with varying degrees of difficulty. To have a setting where children are immersed in science and math, we need to follow the children's interests and what is meaningful to them. Children can best accommodate new knowledge when it is built on existing knowledge. So a study of snow, while it is fun to study and meaningful in areas that have snow, will have little meaning to a child who has never seen or played in the white stuff that falls from the sky. Let me share an experience with you on a "snowy" topic.

When I moved to San Antonio from Brownsville, I thought I was far enough north to experience snow and seasonal change. I could not wait to do "winter" activities with my class. One of my activities used snow! I introduced "snow" to my four-year-old class by using Styrofoam that I had flaked in a blender. Since it snows about once a decade in San Antonio, the children had no idea what snow was, but I was so excited I ignored my own rule: make it meaningful **to the children**. I would later regret it. At the close of the snow study, I talked with the children about what they knew about snow. "White crumbly stuff," said one child. "Yea," said another, "white...and it smells like plastic or something." And what was the biggest and most fun thing about snow to the children? It stuck to their clothes! No child mentioned that snow was cold or that it

could be picked up from the ground, rolled into a ball and thrown at friends in deliriously happy snowball fights. As I stood there reflecting on how I had really blown it with the class, I prayed for a quick weather change. I wanted to erase the "foundation" of smelly, crumbly, plastic snow and replace it with the real thing. That was not a happy day.

Since that experience, I use a study of ice, with which the children are familiar, to provide a foundation for learning about winter and snow. All the activities are based on the concept of cold and hot. When they do play in the snow for the first time, they will have learned about ice, and about cold, hot, freezing and melting. They will have a good foundation on which to enjoy their first snow experience.

Modeling

Another method used in early childhood settings is **modeling**. Children are great imitators, learning to do what they see. If teachers, parents, caregivers and their peers model behavior using science and math in their everyday lives, the children will do the same. When adults demonstrate curiosity about the world around them, the children will "try on" the behavior to see how it is to be curious. If their interpersonal world includes people who have patience with long-term projects, the children learn to be patient waiting for results.

I like to carry a tape measure in my pocket in case something needs to be measured during the day. I will take it out at different times and measure how much a plant has grown or to measure a child's block structure. (I keep a basket of tape measures on the shelf in the Block Center.) Over time I noticed different children with tape measures in their back pockets, so heavy the tapes almost pulled their pants down to their knees. There they were though, measuring all kinds of things, like their head size, the length of their feet, the height of their friends and classmates, how far they jumped and many other things. As time goes by we measure many, many things—with standard and nonstandard measures—and the children begin to understand measurement.

More about modeling. Another time I noticed a child in the Home-living area with a timer in her hand. When another child asked her what she was doing, she said, "I am timing eggs...they cook a long time...2, 20, 5 minutes." She had watched her mom cooking eggs and using a timer. She was modeling her mom's use of time. Modeling is an excellent teaching method.

Questioning and Thinking

Another method is **questioning**. Rather than dispensing knowledge, the teacher needs to ask questions so the children build their own knowledge. It is a far more effective teaching method because the child is part of getting the information. In addition, the child's involvement requires higher-level thinking skills, important skills to be developed. We want them to be able to grasp the meaning of the activity, to **comprehend** it. For example, if you have a block of ice and a sheet of ice, you might ask, "How are they the same?" This lets you know if they comprehend the properties of ice.

We want the children to be able to use learned material in new and concrete situations and to demonstrate the **application** of concepts to other ideas or subjects. For example, you might ask, "How could you use the ice?" Children demonstrate the understanding of application when they demonstrate, solve problems and modify

situations—when they can transfer and use information in other situations.

We also want the children to be able to break down the material into its components so they understand that wholes have parts and that understanding the parts may lead to a more accurate understanding of the whole. This is **analysis.** When children use analytic thinking, they estimate, order, differentiate or separate. For example, you might ask, "What is ice made of? What other things are ice? How much water did it take to make the ice?" This lets you know that the children understand the component parts. If children answer, "I don't know," simply ask, "How do you think you could find out?" This encourages them to think of answers, not look to others for them. It also encourages them to think analytically.

We want the children to be able to put the parts back together to form a new whole. This is **synthesis.** Synthesis engages thinking processes that involve combining, creating, designing, constructing and rearranging. For example, you might ask the children to pretend they are an ice cube and ask, "What would you look like? What would you do? Where would you be? Can you think of another way to make ice?" The answers to these questions tell us if a child can create something new with the information learned.

Last, we want the children to be able to **evaluate**. For example, we might ask, "How does ice help people? How does ice cause problems for people? Do you like ice? Why?"

I am not suggesting that we bombard children with questions. I am suggesting, however, that we catch them at teachable moments and ask them a question or two. When you do this throughout the entire school year, you are asking the children to think scientifically and mathematically.

Acceptance

Because we know that children come to the early childhood settings with different knowledge bases, we need to accept them where they are. **Acceptance** is among the most important methods a teacher can use. When we accept children where they are and accept the efforts they make, we enable them to move forward. We set them up for success. When children feel successful, they learn well.

The Importance of Process

In this context, it is important as teachers of young children that we focus on the process taking place in the classroom, not the content. When we talk about process, we ignore "right" answers and "correct" solutions. We are looking at the methods children use to arrive at some answer. As an illustration, if Susan and Matt are running a race, we are not looking for who is the winner; instead, we want them to learn to run.

Expecting Success

Finally, use the idea of **expectancy** in the classroom. It is magic. When we expect children to become literate in science and math, they will do it. When we say, "What information (research) will you use to prove that?" the children may initially be confused, but when you go for a walk to "collect your data" or you give them a piece of paper to "graph the findings" or if you ask them to "check their source(s)" or you ask them "what measurement did you use," the children will, over time, expect that they ought to do it. More importantly, when expectation is coupled with acceptance, they feel confident enough to do it. They will start to use science and math language and get involved in the scientific approach.

Everyday Discoveries helps teachers blend all of these methods together to create a setting where important learning takes place, and where children are self-directed and in charge of many of their own experiences.

Foundation Materials

The vast majority of the activities in *Everyday Discoveries* use recycled and home-made materials. Some components must be purchased, but not many. All of the activities have been tried. They work. The range of materials teachers typically use could fill a book: from soft-drink, six-pack rings to tree bark, and from yogurt-cup lids to coffee cans. Teachers of young children use anything that enhances the methods they use to teach, and the methods they use are central to building foundations in science and math.

Science and Math Principles

The following science and math principles are the building blocks of learning about science and math. They may sound hard for young children to learn, but I have not found this to be so if they are learning them in a meaningful context and we understand that it is just a beginning. In reading and writing, for instance, we are not interested in mastery, but in children getting a start. The same is true here. The activities in *Everyday Discoveries* that support the following science and math principles are simple and developmentally appropriate. Teachers may discover that they already are doing many constructive science and math activities; they may just not be aware that the skills and concepts they are teaching are science and math related. Each activity focuses on one of the following scientific principles. Since each principle has to be measured to be considered a valid principle, accurate measurement (math) is what suggests that we believe the research and the findings.

Cause and effect

The natural world usually behaves in predictable ways. What happens has an explanation. As teachers of young children, we can help children search for cause and effect and the explanation of it. This is a major activity of science. It lends itself to "if (given) this, then ___?" statements: "If (I hit the rock in a sock with a hammer), then it breaks. I know this because the rock is in a number of pieces and not whole." The child broke the rock and is able to measure it because of the part/whole relationship (pieces) of the rock.

Change

Everything in the natural world is in the process of becoming different than its current form. Often, it is measured by units of time, temperature, or amount—sometimes, all three. "The leaf will change when I put it in the compost aquarium. As it rots, I can see and I can measure the change through observation." The child is measuring the time it takes for the leaf to rot (change), and how much it rots over the time observed.

Cycle

Certain events happen over and over again. These cycles appear not to have an end, but we can know the frequency of the cycle (that is, how often it happens). An understandable cycle for young children: "we watched a frog egg hatch into a tadpole; the tadpole grew into a frog. One cycle took 21 days. When we let the frog go it will lay eggs and the life cycle will continue."

Diversity

There are so many "things" in the world, all different and unique in their own way. Recognizing likenesses and differences is an essential math skill. Example: "There are many different insects on our playground. I know because I can see the differences in each kind." The foundations of science and math rest on seeing the similarities and differences in things. The degree to which they are alike, or different, is a measurement. The number of things alike encourages counting.

Energy

Energy flow is the transfer of force from a source to a receiver. Measuring energy transfer lets us know that energy is moving, by some known or unknown amount. For example, a child realizes that "to make my tricycle move, I have to push with my legs real hard." It is not necessary to know how much (how many pounds of pressure) is "real hard" for the child to begin to understand the concept of energy transfer.

Interaction

Interaction is an event in which two or more objects influence each other. The pattern of interaction can be observed and measured by the degree of interaction. A child realizes that when "I put the magnet next to the paper clips, the clips move. They move very close and stick to the magnet every time." A child is measuring distance and degree of interaction.

Interdependence

Communities, populations, environments and organisms rely on one another to support their basic needs. This is measured by the degree of interdependence. "I water the bean plant. It grows beans and gives me beans to eat. The plant needs me. I need the plant." How the need is measured is math.

Model

A model is a simplified imitation or a representation of a structure or an event. Models suggest how things might work or what they represent. A model is often a "shorthand" symbol, or a convenient abbreviation, of a thing. Models can lead to understanding that an object stands for something, like a number stands for a particular quantity. A child notes, "I ate two pieces of apple." The number two is not two apples, but a representation of them.

Population

A population is a group of organisms of the same kind that live and reproduce in a particular area. Size is determined by the number of individuals in the group: "We saw lots and lots of ants at the ant hill."

Patterns

Objects and events have relationships to one another and they can be arranged as to form, direction or time. Math is the study of patterns. "I can see lots of circles (patterns) in the half (of an) onion and in half a cabbage."

Properties

Properties are qualities that serve to describe or define an object, material or relationship. Color, shape, size, smell and texture are common properties. "I can sort the rocks by shape or by size."

Scale

A scale refers to qualities and quantities. Objects can be identified and arranged by qualities such as texture, shape or size—like seriating sandpaper by the degree of roughness. Thermometers, rulers, magnifiers and scales (weighing devices) are tools for measuring the differences in amount. The child places the pine cones on the table in order by size (largest to smallest), or weighs jars containing different liquid amounts and places them in order by weight.

System

A system is a group of related objects that interact to form a whole. The solar system, a railroad system and a highway system are good examples. A system that is meaningful to young children is a pulley system used, for example, to get wax paper from the bottom of a playground slide to the top (the slide is more slippery and faster if they slide down on wax paper). The children loop the rope around the hand railings and pull up the wax paper in the bucket. Another, perhaps simpler, system is to have mailboxes for each child. Put out a basket of envelopes, paper, canceled stamps (the kind you get in the mail for records, tape and CDs) and pencils and markers for writing letters. Create one more part to this communication system by adding a post office area where the mail has to go first to be canceled using a stamp and pad, and then sorted for delivery.

Science and Math Skills

Young children learn by doing, experimenting, mixing, interacting with their world. They need to practice skills if they are to attain a sense of mastery. This does not mean skill and drill. The children do it themselves when they want to learn something useful or meaningful to them. Let me explain. If you have ever seen a child who has learned to tie a shoe, you will know that children practice (drill) themselves. A child who has learned to tie will tie his shoe, his neighbor's shoe, and every shoe in the classroom if given the chance. Anything that has two strings becomes a bow (or a knot).

Teachers can take advantage of this motivation to learn by setting up centers that reflect the needs and interests of the children. Then it is likely that children will come upon something that interests them as they move through several center areas. When that happens, they will practice and practice, and acquire the skills that are embedded in the activity. For example, when they are tying shoes, they are developing and refining fine motor skills and watching how others do it (observing). If we ask them to tie something else that is less meaningful to them, they will not practice and they will be less inclined to observe. Giving children time to practice skills is critical. They may wish to repeat the activity many, many times to get the full benefit from it. So allowing time for exploration and experimentation is critical. The following skills build a successful foundation in science and math; one or more is developed by each activity in *Everyday Discoveries*.

Observing
This means using the senses to gather information by seeing, feeling, hearing, smelling or tasting, or by using the senses in combination. For example, adopt a tree. The children observe a tree by looking at it, feeling the bark, examining a leaf with a magnifying glass and listening to the trunk with a stethoscope.

Communicating
Communicating is the ability to convey, in some way, information learned, such as creating a graph, drawing a picture or describing the event or information, in groups or individually. This is the beginning of communicating scientific data. Later on in their education, children will make tables, formulate definitions and research information sources. For example, adopt a

tree. The children talk about the tree. They can draw a picture of it or make a graph of the leaves they collect from it.

Comparing
This skill is the ability to look at things and compare them with other similar or dissimilar things. They can do this by estimating, making numerical comparisons or measuring length, temperature, weight, volume, area or time. For example, adopt a tree. The children can estimate the number of leaves and branches, the height or the age of the tree. They can measure the leaves and compare them to the leaves on other trees. They can measure the distance around the tree by holding hands around the tree, by using a string or by measuring with a tape measure. They can compare the size, shape and height of the tree to other trees, and they can compare the shadow it casts at different times of the day.

Organizing
Organizing is the ability to take information and organize it by sequencing, seriating, ordering, sorting, matching, grouping or classifying it. For example, adopt a tree. The children take the leaves and put them in order by size. They can describe the sequence of events required to put them in order by size. They can match leaves or branches, by length, size or shape. They can classify bark from the trunk, from a branch or from an area where the tree has branched. They can group the leaves by color. If you take photographs of the sequence of the shadows at different times of the day, the children can put the pictures in time-sequence order.

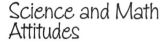

Science and Math Attitudes

Comfort
Children need to feel at ease with asking questions and exploring possibilities. They also need to feel that science and math are part of their everyday lives.

Confidence
Children need to feel they **can do it**; that is, if "it" is not too difficult and beyond their developmental abilities. Confidence also tells them to trust the results of the process, procedure or method they employ. This confidence is often the result of children repeating the same activity over and over.

Cooperation
Children need to learn to work together and to share information and ideas.

Curiosity
Children are born curious, and their curiosity motivates them to find out why things happen, how the world works, to investigate a problem. Nurturing this attitude helps develop children (and adults!) with inquiring minds.

Honesty
Children can use the information they have gathered to prove a point, but truthfully and not distorted to fit their preconceptions.

Open-mindedness
This incorporates many attributes, including the ability to change one's mind and the ability to think about many possible outcomes. Often, once children are committed to a particular idea or outcome, it is difficult for them to change their minds. The ability to envision many possible outcomes is essential to science and math since the vast majority of science is revised, or refined, after further research. Findings are tentative and provisional, subject to change based upon new facts or evidence.

Patience
Some procedures, processes or activities take time. Being able to wait for results (delaying gratification) is one of the most important qualities successful children develop. Children want to find out why and how things happen, and this helps them develop the capacity to be patient. The "finding out" becomes the reward.

Respect
Children need to respect the materials and equipment with which they work. They also must respect living things, and the environment in which living things live.

Tying Science and Math Together With the Scientific Method

Using the scientific method with young children is the best way to help them learn the principles, skills and attitudes that have been discussed above. For young children, a simplified version works best. The following is an example of a group project to discover which color birds like the best. Many of the activities in *Everyday Discoveries* lend themselves to using the scientific method, and each topic ends with an activity that uses the scientific method.

Example
In the spring, we talked about birds and bird's nests. We had found a nest that had fallen from a tree and broken apart, and we were examining all the nest parts. We found many pieces of yarn and several pieces of string, along

with twigs, grass and leaves. The children were fascinated with how the birds got the yarn and the string. We brainstormed several ideas. We tried to figure out why the birds would want the string and the yarn. After we came up with reasons a bird might like to have the yarn in its nest, Megan asked Carlos, "What color do birds like the best?" Carlos had no idea and neither did I. After some discussion, Megan decided she wanted to put yarn in her yard for the birds, and she only wanted to use their favorite color. That was the start of our research! The following is our modified version of the scientific method. Notice all of the math we used to gather our information, as well as the number of science and math skills and attitudes the children were developing.

Statement of the Problem or Question
We wanted to know the bird's favorite color. So our questions was, "What color yarn do birds like in their nests: red, yellow or blue?" Think of this in relationship to the thinking skills you want to develop. This is a simple statement of a problem to solve.

▲ Science and math principle: Teaches about the interaction of birds to yarn.
▲ Science and math attitude: Encourages children's curiosity about the color preference of the birds.

Hypothesis
Each child predicted which color they thought the birds would like the best. We made individual hypotheses and then decided as a group that the birds would like the color blue the best. This process took the children to the application level, where they try to apply their old knowledge to a new problem to solve.

▲ Science and math skill: Encourages children to communicate by formulating a hypothesis and graphing their hypothesis.

9	Thomas		
8	Travis R.	Derrick	
7	Edward	Jennifer	
6	Travis E.	Travis W.	
5	Ashley	Cammi	Carlos
4	Fernando	Ginger	Amy
3	Jeremy	Josie	Victor
2	Jason	Marisa	Pancho
1	Eric	Bridgette	Brittany
	BLUE	YELLOW	RED

Method of Research (testing the hypothesis)
We hung ten pieces of each color of yarn (red, yellow and blue) in a piece of mesh that was stretched over a coat hanger. (Note: turkeys and plastic eggs come wrapped in mesh that can be stretched over a coat hanger and used for this activity.) We hung the coat hanger outside in a tree and checked it each day to see which colors the birds had taken. On our class calendar, we recorded the number of each color of yarn that remained on the coat hanger each day. We continued the observation until all of **one** color was gone (it took more than two weeks for all of the red yarn to disappear, so we developed some patience while waiting for results). Testing the hypothesis leads the children to the analysis level of thinking skills. The children have to analyze how to figure out answers from the events as they observe them.

▲ Science and math skill: Encourages children to observe (and count) the missing yarn.
▲ Science and math attitude: Helps children develop patience.

Checking the Hypothesis

We looked at our graph and at our individual and class predictions to see if we were correct. This process represents synthesis: how do I rearrange the information I have just learned? How do I make it fit into what I know? How are all the parts reassembled to make the whole?

▲ Science and math skill: Teaches children to compare guesses and measure the number of correct guesses.
▲ Science and math attitude: Encourages children to have an open mind and to be honest about the results of the experiment.

Results

The birds chose red yarn. We came up with the result that birds like red the best (an evaluation). What do we think about the results? Why is it valuable information? What is changed because of it? How does it effect us in the long term?

▲ Science and math skill: Encourages children to organize information for future use.
▲ Science and math attitude: Encourages children to keep an open mind.

As a result of our research each child made a yarn hanger for their birds at home. They used a lot of red yarn! They felt they had happier birds living in their red yarn nests in the trees around their houses.

As you go through *Everyday Discoveries* you will see that many of the activities can be done in this format. It does not have to be as formal as this one, but with each activity, the essential processes can be built in. Remember that the children are learning a way to investigate **any problem** they might want to explore, so the emphasis needs to be on the **process** and the **steps** involved. Getting "right" answers is not the purpose. Leave a lot of room for the **children** to be the chief investigators. You will be laying a solid foundation in science and math with this approach. Stand back and watch the learning happen.

Chapter 2

Weaving Children's Interests & Abilities into a Science & Math Curriculum

Over the years, I have learned many things from the children in my classroom that have helped me grow personally and professionally. Children have taught me what **they** know and have shown me, through their actions, when they are ready to learn new things. By observing how children respond in a particular setting, for example, I have discovered what activities, tasks and materials to offer next. This kind of observation and follow-up planning have been called different names over the years, but labels are not important. What does matter is **process.** What is the process? The process is observing and assessing what children are doing, listening to their expressed interests and then using this information to plan the next activity.

Often things happen in the classroom that suggest how to engage young children's attention over a sufficient time period so they can learn important concepts and skills. These "events," for lack of a better word, suggest what the children are interested in already, or interested in now. "Now" is very important. As any experienced teacher knows, **now** is an eternity to a young child and a "teachable moment" to a teacher. I have come to look for these "now" events to show me what the children are ready to do, since young children seldom have the attention span to do things that they are not interested in. Also, these events seem to unify the class. They become interested together, energized by the discovery of the moment. They learn from each other, too.

Much of what they do is "first time" stuff. Sometimes it is hard for children to contain their enthusiasm and curiosity. Take advantage of that. When they are enthusiastic, almost everything is teachable, or so it seems. I enjoy shar-

ing those moments with children. In taking my lead from them, I have discovered various ways to teach a concept or skill. I have found that teaching works best if I take the time to find out what they are interested in and capable of doing. Consider the following story.

The Background

It was early fall. The school year had just begun. We were talking about how the weather was changing, how we had to get ready for winter, what happens outside in the fall and what kinds of changes take place. This was going to lead us, I thought, into a study of all the things that people, animals and plants must do to get ready for winter.

One of the activities I had going in the Art Center was drawing with crayons on white paper that was taped to the inside of one of the classroom windows. On this particular day, the sun shone brightly through the glass. The light was different and we talked a little about how it was different. It was warm. The children drew on the paper with different colored crayons and left the activity as their interest waned. The crayon box remained on the windowsill. I didn't notice it. In my hurry to go home on Friday, I didn't notice that another box of crayons was left on the top of the small classroom refrigerator. It was a short, fat refrigerator. The children loved it because they could see everything inside when they opened the door. Evidently, I was having trouble that day seeing the outside! But after all, it was Friday. Before leaving, I took a quick look around the classroom. I checked the Art Center basket to make sure we were ready for a study of fall on Monday. There were no crayons, so I pulled a box from the closet and set them in the basket. I walked out the door for the weekend.

The Event

During center time on Monday, Lawrence, the child who always was the first to discover new things, noticed the box of crayons on the windowsill that had not been returned to the art basket. Quite a crowd gathered around him as he tried to open the box. He was having a hard time. The crayons inside the box had melted. Several children were offering to help, but everything was stuck together. They could not open the box. No one seemed to know where to begin. Lawrence persisted. He peeled the paper box away from the crayons to find a block of half-melted crayons individually wrapped in gooey paper! Lawrence looked puzzled. So did the rest of the children. The whole class was fixated. They started asking each other questions. No one had any answers except, "Gosh," and "Yea."

Meanwhile, LaKisha had noticed the box of crayons that had been left on top of the refrigerator. She examined it. The crayons inside had melted only a little. She was able to open the box and pull out the contents easily. The partially melted block broke into four clumps as she carried it over to show Lawrence. Yup, this was shaping up as an event, all right!

Building on the Event

My Monday-morning mind raced to help the children use the interest generated by their discovery. I suddenly remembered the box of crayons I had put in the Art Center on Friday. I decided quickly that the children only had two-thirds of an event. I would add the other one-third: a box of unmelted crayons. I pulled them out, hurried over to the children and asked breathlessly, "What happened to the crayons?" They were amazed that my crayons had not changed at all.

We began by talking about what had happened. We decided we needed to name the three boxes of crayons. We called the crayons from the window the "window crayons"; the crayons found on the refrigerator the "refrig crayons"; and the crayons from the Art Center the "just plain crayons." The children described the differences in the three boxes of crayons. We made a long list of differences (Figure 1). We discussed where each box had been found.

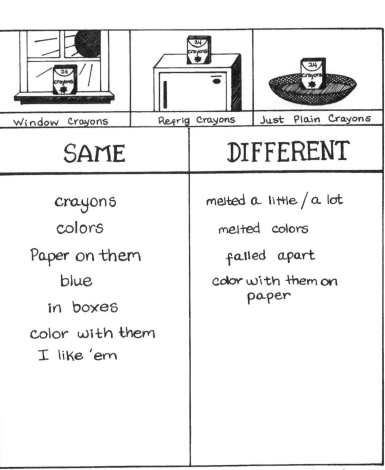

| Window Crayons | Refrig Crayons | Just Plain Crayons |

SAME	DIFFERENT
crayons	melted a little / a lot
colors	melted colors
Paper on them	falled apart
blue	color with them on paper
in boxes	
color with them	
I like 'em	

figure 1

The crayons melted in the window because...

- they were in the window
- they got hot like my car window
- they sat there

The crayons melted on the refrigerator because...

- it was hot

The crayons did not melt in the art basket because...

- it was hot
- it was not hot
- it was cold
- they were in the box

figure 2

The children guessed why they thought the crayons had melted. We made a list of those guesses (Figure 2). They examined the crayons further by peeling back the paper wrappers and by coloring with blocks of melted crayons on sheets of white paper. They broke the crayons to see if the inside and the outside were alike. So it went the whole day.

Finding Out What the Children Knew

During group time on Tuesday, we developed a "Circle of Knowledge" (Figure 3). As we discussed what the children knew, I recorded their responses to four topics in this order: Crayons, Colors, Different Colors, and then Color Mixing. Please note that their responses do not make much

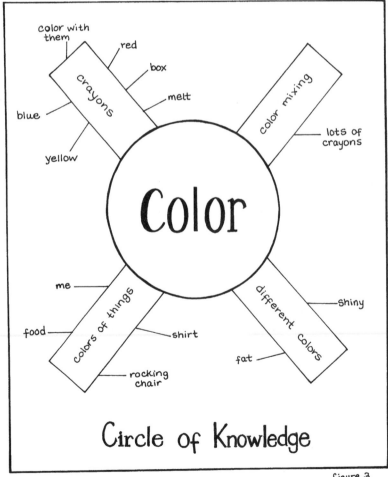

Circle of Knowledge

figure 3

sense (from an adult perspective) as they moved from the subject of something they knew much about (crayons) to a subject they knew little about (colors of things, different colors and color mixing). The purpose of the Circle of Knowledge, then, was to help me know what they knew, and to help them focus on the familiar but stretch to the less familiar. I phrased questions like: "What do you know about crayons?" They knew quite a lot as you can see (Figure 3). The next thing I asked them was, "Tell me about the color of things." Here they had some knowledge but not as much as about crayons. When I said, "Tell me about different colors," they were stumped. Last, I asked about color mixing. They had virtually no fund of knowledge on color mixing. The Circle of Knowledge became my **planning guide** for what do now, and what to do next. Tuesday was spent.

On Wednesday, I asked the children to think of the things they could do with crayons to learn more about them. They generated a list of all possible activities (Figure 4). This was how I learned from them what they wanted to do next. Figures 1, 2, 3 and 4 were posted on the wall for the children to refer to if they wished; information about what they learned as they learned it was added as they went along. This helped the children look back on their earlier fund of knowledge and realize that they were adding to it. Thursday and Friday were holidays. The time gave me a chance to think about the past three days and to plan what I was going to do when we returned to school.

The Plan

This observation and assessment of the children's interests and abilities led me to shift from the topic of fall to that of

crayons and colors. Why? Because that's what the children were interested in and motivated to learn about. One of my goals as a teacher of young children is to teach concepts and skills, but I must recognize their developmental level, teach to their interests and use their enthusiasm whenever possible. I decided to let them explore colors and crayons for as long as their interest lasted, knowing they would learn many concepts and skills even though I had abruptly changed topics.

How to begin? I asked myself: What are the underlying science and math principles embedded in the study of colors? (Please see Scientific Principles, beginning on page 16.) Well, one of the principles could be **diversity**. There are literally thousands of colors and many types and sizes of crayons. On the other hand, I could have chosen **interaction,** since many interesting things happen when heat acts on

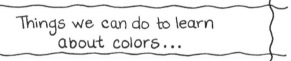

Things we can do to learn about colors...

- color with them
- smell them
- hold them
- count them in boxes
- melt them

figure 4

crayons, and crayons act on paper. I decided, however, to focus on the **properties** of colors. By properties, I mean *characteristics* (i.e., what is it that makes something what it is?). This would provide a good foundation on which to build the understanding of diversity and interaction. At the start of the study, I could use colors and crayons

TOPIC: Colors

THEME: properties

Music Center
- colored streamers
- chiffon scarves
- Hap Palmer: Colors
- Bev Bos: World Outside My Window
- Greg and Steve: Rainbow Song

Block Center
- add crayon candles
- add colored fabric pieces
- colored blocks
- colored flags with bases
- colored house signs and numbers
- colored windows
- cellophane drapes

Construction Center
- Leggos
- painted PVC pipes
- painted cardboard slots

Outdoor Center
- cellophane streamers
- stocking ball
- stocking bat with pompom
- sunshine rainbow

Woodworking Center
- add colored sandpaper
- colored pieces of wood
- colored nails
- colored paper shapes for nailing on construction
- paint construction
- colored telephone wire for decorating construction

Science Center
- prism
- color mixing
- weigh crayons
- seriate crayons
- crayon graph
- crayon predictions
- Crayon patterns
- color paddles
- color mixing bottles

Library Center
- Multiplentication poem
- crayon shaped blank book
- books about colors
- color words box
- overlay color books
- overhead: colored shapes
- Scat the Cat flannel board story

Pouring Center
- colored water
- raindrop race
- colored sand
- shaving cream with food coloring
- colored ice melting
- colored baggies and water

Art Center
- crayons on sandpaper
- crayon melt
- record player art
- eye dropper art
- crayon cookies
- tissue collage
- water colors
- broken crayon collage

Manipulative Center
- crayon sort
- color gradation
- color bingo
- crayon match
- measure with crayons
- crayon box puzzle
- color mixing nesting boxes
- colored popcorn sort

Dramatic Play Center
- color goggles
- color survey
- colored clothes
- color pendants
- colored plates, flatware, cups and napkins

figure 5

virtually interchangeably; to the children, crayons were colors (even though they are a **representation** of color).

I used a classroom event that engaged the children's interest, and I used their existing fund of knowledge as a starting point from which they could learn something new. The subjects were tied together, somewhat arbitrarily by me, but they occurred in a logical progression from crayons to colors of things, to different colors and to color mixing. Each step was built on the last. This would be the order I would use to focus the children's interests and abilities each day. I would start with the familiar and work toward the unfamiliar.

The next task for me was to decide which **concepts** I wanted the children to take from the study. I came up with these:

✓ There are lots of different colored crayons.
✓ Crayons contain colors that we can use like tools.
✓ Colors help us see the similarities and differences in things.
✓ Colors can be mixed to make new colors.

What about science and math **skills**? The children would use or develop many skills during the study of crayons and colors. These included observing, communicating, comparing and organizing. (See pages 18-19 for a breakdown of each of the science and math skills.) In addition, I tried to add new classroom activities—as many as I could think of—to punctuate and build on what the children already knew about crayons, colors of things, different colors and color mixing. I brought in the principle, reinforced the concepts and then offered opportunities for new skills to be learned.

My goal was to immerse the children in colors, not just in the Science Center but in **all the centers** throughout the classroom. You can see what I came up with in Figure 5. I wanted to be sure that the properties of colors were embedded in every activity in the classroom. This reinforced the science and math principle being taught. I incorporated the children's ideas, too, as they came up with them. I knew some of my activities would not engage the interests of the children (they would "bomb"). That is just the way it is. But some would be hits. I also knew that I could not use all of the activities at once, so I would add new ones as the children showed interest and subtract others when necessary. The success of this teaching approach is possible by observing and assessing the children's interests and abilities and using their delightful enthusiasm for new experiences.

Following the Plan

Let's look at the first week of the study of colors by examining what was done during morning group. I began the week by looking at different kinds of crayons in boxes. To get the children started, I used morning group time to focus on aspects of crayons and colors. We talked about similarities and differences. We compared, for example, Prang crayons, Crayola crayons, glitter crayons, crayons in gem tones, jumbo crayons, crayons that glow and chunky crayons. I divided the children into groups of three or four. Each group explored two or three boxes of crayons. During morning group time, the children dictated how the crayons were alike and different. I recorded their observations on chart tablet paper (Figure 6).

On Tuesday we focused on one box of twenty-four crayons and, with much debate, selected the colors that fit in categories of blue, yellow, red and other, meaning those that were left over (Figure 7).

Wednesday and Thursday group time was spent predicting, counting and weighing crayons (Figures 8 and 9). On Friday we began the next part of our study: the colors of things. During group time, we read *Red Is Best* by Kathy Stinson. This story tells about a little girl who gave every reason she could think of for why red is the best color. For example, she preferred red stockings to white, her red jacket to her blue one and her red boots to any others since she knew she could take longer steps in red boots! The book has many other examples of why red is best. The children enjoyed it. Afterward, we looked at all the red things the book said she preferred. We discussed and dictated why we thought she preferred them. As a part of this activity, the children each selected a favorite color and gave reasons why (Figure 10). After the first week, the foundation had been laid for moving to the next part of the study: the colors of things. The activities and materials were still available for exploration by the children in centers, but my focus in morning group was on learning about the color of things.

Keeping the Interest High

To keep interest at a high level and to keep the children enthused, I try to get the parent(s) involved. Parental interest fuels the children's. I do this by transcribing the morning group chart, making copies and sending them home. I also send a short note home so parents know what's going on at school and how they can help reinforce what is being learned (Figure 11).

This study was naturally concluded when the children lost interest. We then moved to another topic. The

SEPTEMBER 23

Today is Monday.
Victor is our leader.
How are the boxes of crayons

__alike__	and	__different__
colors		fat and skinny
red		colors
paper 'round 'em		big and little box
color on paper		long and tiny
		1 - no paper

figure 6

SEPTEMBER 24

Today is Tuesday.
Cammi is our leader.
How many colors are alike in a box of 24?

		more
red	I I I I I I I I	⑧
yellow	I I I I	4
blue	I I I I I	5
other	I I I I I I I	7

figure 7

exciting thing about these "events" is that they can be revisited over and over throughout the year. As other subjects were introduced, I repeated some of the activities developed for crayons and colors, as dictated by observing and assessing the interests and abilities of the children. They loved the repetition and I enjoyed seeing their fund of knowledge about science and math expand.

My story stops here. I hope your story begins with a few new insights about how to build on the children's natural interest in new things and how to tap their natural enthusiasm.

SEPTEMBER 25

Today is Wednesday.
Carlos is our leader.
We can guess how many crayons
are in the jar.

Carlos 9	A.J. 2	cammi 26	Robert 14
Latron 5	marissa 10	Jeremy 222	victor 14
megan (25)	Jose (25)	Thomas 3	scotty 2000
Eric 2	Denise 21*	Keisha 10	christopher 10
Vanessa 9	Tom 11	Tabitha 12	Amy 10
Omi 12	Kei 10		

(25)

figure 8

SEPTEMBER 26

Today is Thursday.
Latron is our leader.
How many chips does it take to balance the crayon in the balance scale?

1 crayon	__1__ chip	25 – 25
2 crayons	__2__ chips	50 – 50
3 crayons	__3__ chips	100 – 100

figure 9

SEPTEMBER 27

Today is Friday.
Marissa is our leader.
Red is Best by Kathy Stinson
What name could we give the little girl?
        Ariel
Why do you think she liked red best?
    happy   loud   fun   cold
What is your favorite color?

yellow |||||| red ||||||||| blue |||| green ||

figure 10

Dear Parents,

We have had a fun week learning about crayons. We learned how crayons are alike and how they are different. We used crayons in many, many ways. We learned that crayons are tools we use to give color to our work. Next week we will be talking about colors. Please help your child find colorful things in your home. Let him bring one object to school. We will see how many things have different colors and how many are the same color. Understanding same and different is a very important skill for your children.

Your child's teacher,
Sharon MacDonald

figure 11

Chapter 3

Everyday Discoveries: Science & Math Activities

Interest area
Cooking and snack

Science & math principle
Teaches children about how apples change.

Science & math skill
Teaches children to organize information as they follow directions.

Science & math attitude
Encourages children to cooperate when they work together.

▼

You need

10 to 12 apples
apple corer
½ to ¾ cup (125-175 ml) water
½ cup (125 ml) sugar
1 teaspoon (5 ml) cinnamon
crock pot
cup measure
teaspoon measure
cutting board and knife
spoons
cups
napkins

Making Applesauce

Activity

- ▲ Follow the rebus (picture directions) to make the applesauce.
- ▲ Encourage the children to help read the directions and to assist with each step of the process.
- ▲ Serve the applesauce in the cups.

Questions you might ask

Why do you think we made applesauce?
How is this like applesauce bought in the store? different?

Extension

Experiment by making other sauces from fruit like cranberry sauce.

More challenging for older children

Compare the flavor of store-bought applesauce to the applesauce made in class. Make a bar graph to note each child's preference for flavor. Call the graph "The Applesauce We Like Best."

Modifications for younger children

To help the children remember their applesauce-making experience, ask them to describe the sequence of events using the picture directions as a guide.

Core 10 to 12 apples.

cut them into small pieces.

Put them in a crock pot.

Add ½ to ¾ cup of water.

Add ½ cup of Sugar.

Add 1 teaspoon of cinnamon.

Put on the lid. Cook overnight on low heat.

Eat!

Freezing Apples

You need

3 apples
plastic knives
salt
water in a pot
colander
large bowl
$^1/_3$ cup (75 ml) sugar
resealable plastic bags
stove and freezer

Activity

▲ With the children's help, peel the apples with plastic knives.
▲ Rinse the apples and then slice them thinly.
▲ Drop the slices into slightly salty water and allow them to boil for one minute.
Note: Supervise closely, stressing the danger in boiling water.
▲ Drain the apples slices.
▲ Rinse the slices under cold water.
▲ Dump the apple slices in a large bowl.
▲ Sprinkle the apples with the sugar.
▲ When the sugar has dissolved, pack the apples slices in baggies and freeze. Do this at the beginning of your study of apples; eat them for snack at the end.

Questions you might ask

▲ What do you think the boiling salted water did to the apples?
▲ Why do you think we use sugar on the apples?

Extension

Freeze various kinds of fruit in the same manner and observe what happens to each. Compare the results and have the children taste each fruit.

More challenging for older children

Compare unfrozen, sliced apples with frozen apples. List the differences and similarities and talk about them.

Modifications for younger children

Peel, core and wash all the apples but one before the children arrive in class. Peel the last one with the children and proceed with the activity. If you peel the apple the night before, be sure to coat it with lemon juice so it will not turn brown.

Withering Apples
(by the Slice)

Activity

▲ Slice the apple horizontally, in thin sections. Leave the skin on.

▲ Thread the string through the needle (use a large, plastic darning needle so the children can help).

▲ Thread the needle through the center of each apple section.

▲ Space apple slices so they do not touch.

▲ Place the string of apple slices inside the mesh bag or cheesecloth (to keep insects away), tying off any open ends.

▲ Hang it across a sunny window.

▲ Allow the apples to dry (it takes from ten to fourteen days, depending on the weather).

▲ When the apples are dry, eat them for snack.

▲ Compare the taste, texture, look and smell of the dried apple to a fresh, sliced apple.

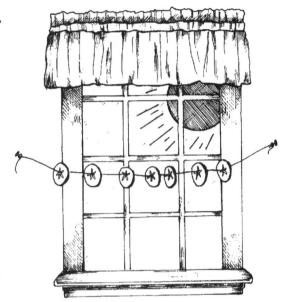

Hang the apples in front of a window that gets sunlight.

Interest area
Cooking and snack

Science & math principle
Teaches children about cause and effect as the sunlight dries the apples.

Science & math skill
Encourages observation skills.

Science & math attitude
Develops patience as children wait for the results.

▼

You need

1 apple
knife and cutting board
plastic darning needle
string
mesh bag or cheesecloth
sunny window

Extension

Dry a whole apple. Core and peel it. Soak it with lemon juice for thirty minutes. Blot it dry with a paper towel. Set it on top of a spool of thread and let it dry for three to four weeks. Compare it to a fresh apple. Brainstorm with the children the reasons why the dried apple was so much smaller in size.

More challenging for older children

Ask the children to predict how many days it will take for the apple slices to dry. Design a graph to display their predictions and check them for accuracy after the apples have dried.

Modifications for younger children

Do this activity in small groups and allow each child to slide a slice of apple on the string. Be sure to place apple slices beyond their reach but in full view when they are drying. Involve the children in checking the slices frequently for dryness. This will help them remember the activity.

Questions you might ask

▲ How else could you dry apple slices?

▲ What is the same about the dried and fresh apple slices? different?

Interest area
Group time

Science & math principle
Teaches about properties of apples.

Science & math skill
Encourages children to communicate what they have learned.

Science & math attitude
Encourages children to be comfortable with science and math.

▼

Bon "Appletite"

Activity

▲ Before the children arrive for class in the morning, set out one paper plate for each child. Put three paper cups on each plate.

▲ Pour a small amount of apple juice into one cup, scoop a little applesauce into the second and place a piece of apple into the third. Do the same for each set of three cups.

▲ When the children arrive, pass out the cups on the plates, along with napkins and spoons.

▲ Explain that apples can be prepared in different ways and that you are going to taste some of those ways. As the children taste the contents of each cup, talk about texture, smell and flavor.

▲ When the children have finished tasting, have them decide which they liked the best (juice, sauce or chunk) and mark it on a graph, if desired.

Note: Save all the apples from the apple activities to make applesauce. Directions for making applesauce appear in "Making Applesauce" on page 34. Apple pieces left over from an activity can be soaked in lemon juice, placed in a plastic bag and refrigerated until you are ready to make the applesauce.

You need

apple juice
applesauce
an apple
cups
napkins
paper plates
spoons

Questions you might ask

▲ What other types of food made from apples have you eaten?

▲ How are applesauce and apple juice alike? different?

Extension

Taste and graph other ways that apples can be prepared (e.g., apple pie, apple dumplings, sun-dried apples, apple cobbler).

More challenging for older children

Put the graph in a station where children move independently. Have them post their own preferences on the graph.

Modifications for younger children

After they have tasted the apple products, gather the children in small groups at centers. Show them the apple products again and help them graph their preferences.

Apple Star

Activity

▲ Tell the children the story below about a boy named Brad who found the apple star. Substitute seeds from your geographic area if you do not have the seeds mentioned in the story.

▲ When you get to the part in the story where the father cuts the apple in half vertically (stem pointing upward), demonstrate by cutting your own apple vertically. Show the children that there is no star.

▲ When you tell how the father cuts the apple in half horizontally (crossways), cut your apple so the children can see the star.

▲ Cut the star apple into small slices so that each child can have a small taste of apple.

Interest area
Group time

Science & math principle
Teaches children about the properties of apples.

Science & math skill
Improves children's observation skills.

Science & math attitude
Encourages children's curiosity about apples and their characteristics.

▼

You need

2 apples
1 knife
1 tray

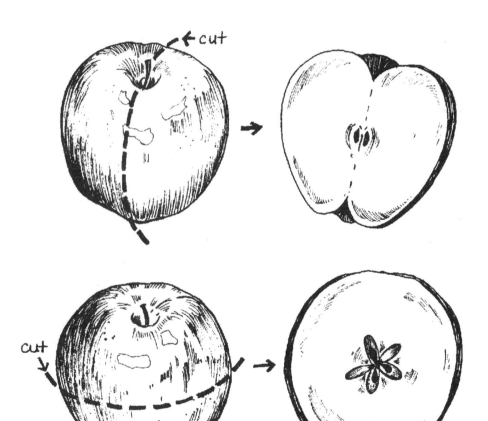

THE APPLE STAR
by Sharon MacDonald

Once there was a little boy named Brad. He lived on a farm where there were many kinds of trees.

Brad was a very curious little boy. He liked to climb and explore in the trees. He poked in every hole and peeked in all the birds' nests. He collected acorns from oak trees and pine cones from pine trees. Brad ate the apples from the apple trees, and he made helicopters from the maple-tree seeds. He liked the apple trees and apples best of all.

One day, his daddy told him that if he looked carefully he could find a star hidden in one of the fruits or seeds from one of the trees. Brad couldn't wait to find it.

He looked under the cap of the acorn. No star. He took apart the pine cone piece by piece. No star. He peeled the maple seed. It wasn't there, either.

Brad was very sad, and he was very tired of looking for the star. He sat down on the porch to think. Soon his daddy came out with two apples and sat down beside him. His daddy pulled out a small pocket knife and began to cut the apple in half, stem up, right down the middle. Brad stopped him and looked at his daddy's apple. No star. Then Brad had an idea!

"Daddy, please cut my apple across the middle." He held his breath while his daddy cut. Brad peeked inside and there it was, the apple star! He was so excited and happy that he had found it. Daddy handed him the other half, and there was a star in that half, too.

"What could be neater than finding a star in an apple?" Brad asked.

"Finding two!" said his daddy.

Questions you might ask

▲ Into how many parts did Brad's daddy cut the apple each time?
▲ How did Brad get two stars?

Extension

Ask the children to make up a story about apples.

More challenging for older children

Ask the children to retell the story by recording it on tape or by drawing pictures that depict the sequence of events.

Modifications for younger children

Pass around acorns, pine cones and maple tree seeds for the children to see and feel when you are telling the story.

Sing a Song of Apples

Activity

▲ Draw a red circle on an index card and color a patch of red on another.
▲ Glue the pictures on separate index cards.
▲ Write individual lines of the song on sentence strips. Put a piece of Velcro above the words that will be replaced by a picture. (Line 1, apples; Line 2, seeds; Line 3, round circle and red patch of color; Line 4, apple trees; Line 5, applesauce; Line 6, apple juice; Line 8, apple pie)
▲ Put a piece of Velcro on the back of each index card.
▲ The first time you sing the song, individual children place the sentence strips in the pocket chart as they are sung.
▲ The second time, children place the picture over the appropriate word as it is heard in the song.
▲ After the children are familiar with the song, place the sentence strips, pocket chart and pictures in the music center for the children to explore.
Note: We do not always think of a song as being the "stuff" of science and math. However, music is mathematical; sound followed by silence is pattern. Math is the study of patterns. If we examine what is happening in a song, we realize that the rhythm is a pattern of beats and that rhyme is a pattern of words. Differentiating volume and listening for sound vibration are all part of music and singing. Each time we sing, we are not only learning and practicing science and math, but we are also internalizing that learning so quickly that we may fail to recognize its importance.

Sing a Song of Apples by Sharon MacDonald

(tune: "Sing a Song of Six Pence")

Sing a song of apples,
Apples full of seeds.
Apples that are round and red
And grow on apple trees.

Apples make great applesauce
And juice that you should try.
My favorite way of eating them
Is in an apple pie!

Interest area
Music and movement

Science & math
principle
Teaches about how apples change.

Science & math
skill
Encourages children to communicate by singing.

Science & math
attitude
Teaches children to work together cooperatively.

You need

markers
index cards
glue
pictures (one each of a
   group of apples,
   an apple tree,
   a jar of applesauce,
   a can of apple juice,
   an apple pie or
   a slice of apple with the
   seeds visible)
apple seeds
pocket chart
sentence strips
hook and soft-side Velcro

apples

Questions you might ask

▲ Without looking at the words, can you name the apple products mentioned in the song?

▲ What other apple products can you name?

Extension

Try to rewrite the song using other fruits.

More challenging for older children

Leave out one word from the sentence strips. Ask the children which picture goes in which line.

Modifications for younger children

Glue the pictures to the song strips.

Too Long, Too Short, Just Right!

Activity

▲ Make a chart (see illustration).

▲ Place the string, the apple and the scissors on the tray below the chart in front of the baskets.

▲ The children pull a length of string from the ball and cut it to the exact length they think they will need to go around the apple (the circumference).

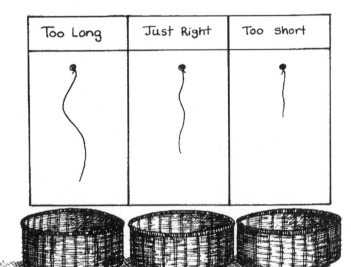

Too Long	Just Right	Too Short

▲ Next, they verify their estimate by putting the string around the apple.

▲ If they guessed correctly, they place their string in the "just right basket"; if it is too long, they place it in the "too long basket"; and, if it is too short, they place it in the "too short basket."

▲ Children make estimations and cut lengths of string as many times as they want.

▲ This is an excellent activity to introduce the word "circumference."

More challenging for older children

Ask children to estimate the circumference of the apple eight to ten times, using a new string and checking it each time. When finished, have them line up their strings in order of their guesses (1, 2, ...10). Notice how their estimations of the apple's circumference become progressively more accurate; that is, experience leads to more accurate estimates.

Modifications for younger children

Show the children three strings of different lengths. Ask them to choose the one they think will go around the apple, then have them check to see if they are right.

Interest area
Science and math

Science & math principle
Teaches about scale (size) as children discover the qualities of apples.

Science & math skills
Teaches children about comparing sizes.

Science & math attitude
Encourages children to be open-minded as they learn by trial and error.

▼

You need

an apple
ball of string
tray
scissors
chart tablet and marker
3 small baskets

Questions you might ask

▲ What else is as big around as an apple?
▲ What else could you use to measure the apple?

Extension

Do the same activity using a pumpkin, a large rock, a can, a ball, a watermelon or a child. Make a string chart of the different lengths so the children can compare them.

Apple in a Square: Why Did You Put it There?

Interest area
Science and math

Science & math principle
Teaches about patterns using different colored apples.

Science & math skill
Encourages children to communicate what they have learned in a graph.

Science & math attitude
Develops children's curiosity about apples.

▼

You need

4 red apples
3 yellow apples
2 green apples
basket to hold all the apples
large piece of paper
marker

Questions you might ask

▲ Why do you think apples come in different colors?
▲ What are other ways to sort and graph the apples?

Extension

Graph the apples by characteristics other than color.

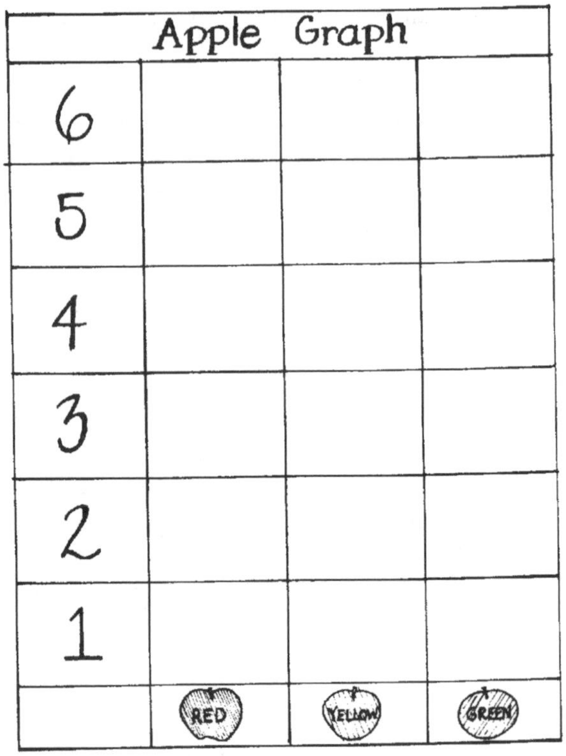

Apple	Graph	
6		
5		
4		
3		
2		
1		
	RED	YELLOW	GREEN

Activity
▲ Make a graphing grid like the one pictured here.
▲ Place the graphing grid on the floor.
▲ The children graph the apples by color on the graphing grid, starting with square one.
▲ Ask the children to explain as they place an apple in a square why they chose to put it there.
▲ Discuss how many of each apple you have on the graph.

More challenging for older children

Give the children a blank grid. Ask them to figure out their own method of graphing.

Modifications for younger children

Color one of the apples at the bottom of the graph red, another yellow and the third green. This gives children a visual clue to help them sort and graph the apples.

Apple Weighing

Activity

▲ Put out a balance scale, a kitchen scale and a container of apples for the children to explore weighing.

▲ The children figure out how many apples it takes to balance the scale.

▲ Then they confirm this information by weighing the apples on the kitchen scale.

Note: In this activity, the children are comparing an equivalent number of apples balancing on the balance scale with the actual numerical weight of the apples when weighed separately on the kitchen scale.

Questions you might ask

▲ Can you think of a new way to weigh the apples?

▲ Which scale do you like the best?

Extension

Weigh rocks, seeds, buttons, crayons, blocks and other items on the two scales, comparing results.

More challenging for older children

Repeat the same activity but make comparisons with different colored (types of) apples. For example, older children can determine which apples have the most water content and are therefore heavier.

Modifications for younger children

Introduce and use the balance scale. Bring in the kitchen scale later, for comparison.

Interest area
Science and math

Science & math principle
Teaches children about the weights of different apples (scale).

Science & math skill
Encourages children to compare the differences in weight.

Science & math attitude
Teaches children to respect tools used to gather information.

You need
kitchen scale
balance scale
6 to 8 apples (the same type)
container

Apple Float

Interest area
Science and math

Science & math principle
Teaches about patterns of how apples float.

Science & math skill
Encourages children to compare their predictions and findings.

Science & math attitude
Develops the ability to be open-minded.

You need

4 or 5 apples, various sizes and types
chart tablet
marker
dishpan
water

Questions you might ask

▲ How did your predictions change after the first apple floated? the second? the third?
▲ What predictions can you make about the behavior of apples when they are placed in water?

Activity

▲ Fill the dishpan half full with water.
▲ Ask the children to predict if an apple will float and, if so, if it will float stem up, stem down or sideways.
▲ The children mark a chart with their predictions, then test their predictions by floating the apple in the tub of water.
▲ They compare the results with their predictions.
▲ Repeat the process until all the apples have been tested.

Extension

Use other fruits, like pears or quinces, to do the same experiment. Compare an apple and pear to fruits like a plum, cherry or peach and to citrus fruits like an orange or grapefruit. Bob for apples.

More challenging for older children

Ask the children how they think they might graph their predictions so they can check their predictions against the results.

Modifications for younger children

Do the same activity without using the graphs. The children just call out their guesses. Discuss what happened to each apple.

Apple Rotting

Activity

Note: This activity can take from three to five months to complete.

▲ Place the apple in the container.

▲ Put it in an out-of-the-way place in the classroom, visible for observation but away from the heavy traffic areas.

▲ **State the problem or question**
How many days will it take for an apple to rot?

▲ **State the hypothesis**
We think it will take ___ days.

▲ Let each child predict what will happen. Make a list of their predictions. This can be a consensus prediction or a collection of individual predictions.

▲ **Method of research**
We will set aside an apple to rot. We will examine it every day to see the changes in the apple and we will talk about the changes. We will put an "X" on the calendar if the apple has not rotted completely by the end of that day. We will wait until the apple is completely dry before we consider the apple rotted. When the apple is rotted, we will end our observation.

▲ **Check the hypothesis**
We will count the days it took for the apple to rot and check it against the predictions. We will count the first and the last day of our observation in calculating the total number of days.

▲ **Results**
It took the apple ___ days to rot.

Extension

Try this with other fruits that represent seasonal change.

More challenging for older children

Ask the children to draw pictures of the apple as it changes. Save the pictures and let each child make a booklet of the rotting apple.

Modifications for younger children

Repeat this activity several times with different fruit until the children understand the rotting process.

Interest area
Science and math

Science & math principle
Teaches about change as the apple rots.

Science & math skill
Develops keen observation skills.

Science & math attitude
Develops patience as the children wait for the results.

You need

1 over-ripe apple
small see-through container
calendar
chart tablet and marker

Questions you might ask

▲ How is the rotten apple different from the apple at the beginning of the experiment?

▲ What do you think caused the apple to rot?

Science & math principle

Teaches about cause and effect when the collage materials stick to the paper.

Science & math skill

Develops observation skills.

Science & math attitude

Encourages children to be comfortable with science and math.

▼

You need

2 or 3 eggs
egg-yolk separator
2 small containers
2 trays
paintbrush
paper
scissors
confetti
magnifying glass

Questions you might ask

▲ How does the egg white look like glue?

▲ Why do you think egg white makes things stick to the paper?

Gluing with Egg White

Activity

▲ Before the children arrive, put the following on a tray: one egg, the egg-yolk separator (a tool used to separate the egg yolk from the egg white), the paintbrush, the small containers, a sheet of paper cut into shape of an oval and a little confetti.

▲ Help the children separate one of the eggs using the separator; drain off the egg white into a container. Talk about part and whole.

▲ Set the yolk aside in a container for the children to examine in the Science and Math Center with a magnifying glass.

▲ Gently stir the egg white with the paintbrush. Add a touch of water if the white is too thick. Ask the children to describe the egg white. Talk about other uses for egg white besides eating it.

▲ Encourage one child to brush the egg white on the sheet of oval paper and another to sprinkle confetti over it.

▲ Ask the children to predict what will happen when you pick up the paper.

▲ Wait a few minutes for the egg white to dry. Pick up the paper and shake off the excess confetti in the other tray. Gently turn over the paper.

▲ Ask the children if their predictions were correct. Talk about how you used egg white as a glue. Mention that some cultures use it as glue.

▲ Place the materials in the Art Center so the children can explore using the egg-white glue.

▲ Use the other eggs to make more glue if you need it.

Extensions

Try making glues from other foods, like potatoes, corn, wheat or rice. These can be boiled down into a thick paste. Mash them continuously while they are cooking, keeping enough water in them to prevent burning. You will end up with a thick, starchy paste that adheres like glue but will hold only very lightweight materials to paper.

More challenging for older children

Let the children do most of the work. Have them write their predictions and check them. Encourage them to try different collage materials.

Modifications for younger children

Do this in a small group or with individuals. Give the children small, stable containers with a small amount of egg white and large brushes to use. Give each child a shake-off tray. Help them work inside the tray by placing the paper in the bottom of the tray. Put a small bag of confetti and the small container of egg white inside the tray.

Tasting Eggs

Activity

▲ Before the children arrive, scramble and hard boil enough eggs to give each child a spoonful or a slice to taste. Keep the scrambled eggs warm until they are ready to serve.

▲ Cut the large sheet of paper into an oval shape and draw on it seven smaller ovals titled:

✓ Scrambled eggs
✓ Fried eggs
✓ Hard-boiled eggs
✓ Scrambled and hard-boiled eggs
✓ Scrambled and fried eggs
✓ Fried and hard-boiled eggs
✓ All three kinds

▲ Discuss with the class the ways they have eaten eggs at home and in restaurants. Talk about the three ways that you will prepare the eggs for tasting.

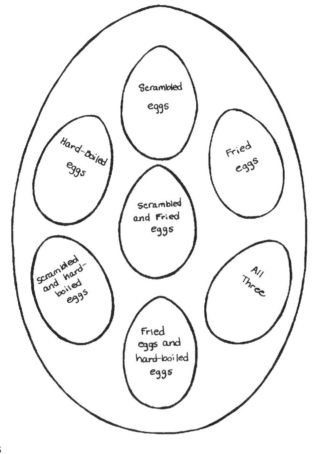

▲ Show them two of the hard-boiled eggs. Let them help you peel them. Show them the egg slicer and demonstrate how it works. Let them slice the remaining hard-boiled eggs. Place slices aside to taste later.

▲ Set up the skillet, the eggs and the butter on a tray. Talk about how the skillet gets very hot and that it can make serious burns if it is touched. Melt a small amount of butter in the skillet and allow it to get hot. Crack an egg and place it carefully in the butter to cook. Fry the egg in the butter on one side and turn it over. When it is done, put it on a plate and cook two or three more, enough for each child to have a taste.

▲ When you have fried the eggs, divide them into small pieces and place them on the paper plates. Talk about how the egg looked when it was whole and how a fried egg looks when it is cut into pieces.

▲ Place some scrambled egg and a slice of hard-boiled egg on each plate. Have an egg-tasting party.

Interest area
Cooking and snack

Science & math principle
Teaches about changes when the egg is fried.

Science & math skill
Encourages children to compare how eggs taste when prepared different ways.

Science & math attitude
Encourages children's curiosity and desire for knowledge.

You need

1 egg per child
egg slicer
electric skillet
cooking utensil
butter
salt
large sheet of white paper
scissors
marker
napkins
paper plates
forks

▲ When all of the children have tasted the eggs, ask them to decide which way they liked their eggs.
▲ When the children choose which they liked the best, write their names in the corresponding oval. After everyone has chosen, count how many children are in each oval.
Note: You might want to make an eighth oval titled "None" in case you have children who do not want to taste any of the eggs.

Questions you might ask

▲ What is the same about your eggs? different?
▲ How many children's names are in the oval with your name?

Extension

Think about other ways to cook eggs and to serve them (poached, omelets).

More challenging for older children

Do the activity over several days. Let the children assist in the cooking. Have them cook a different kind of egg each day and taste it. Since older children can remember their preferences, use the last day to record and review the three cooking processes.

Modifications for younger children

Serve the plates with each type of cooked egg, but hold back one egg to fry with them and one hard-boiled egg to slice in front of them. Spend your time tasting and recording their choices and preferences. If the children get restless, do not record their preferences.

Feather Uses

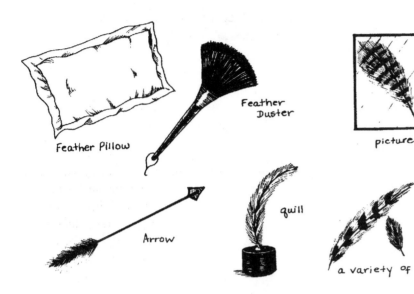

Feather Pillow

Feather Duster

picture

Arrow

quill

a variety of feathers

Activity

▲ Make a small opening in the old feather pillow to remove some of the feathers.
▲ Show the children several kinds of feathers.
▲ Allow them to feel the feathers.
▲ Ask them to tell you what they think people do with feathers. Make a list of all of the children's suggestions on the chart paper.
▲ When they have finished, show them the items listed, one at a time. Talk about how the feather is necessary to each item's function and how people use the items.
▲ Compare the number of feathers in each item, the size of the feathers used, how the feathers are organized on the item and so on. Have the children decide which items they like the best and set out those items in the Science and Math Center for them to explore.
Note: Be sure to close up the opening you made in the feather pillow!

Extension

Ask a bird expert (from a pet shop or a college or university) to come to school and talk about how the birds use the different kinds of feathers on their bodies. Such an event might be coordinated so that other classes in the school could participate.

More challenging for older children

In small groups, have the children state why each item is useful and important to people.

Modifications for younger children

Share the feathered items with the children and let them talk about them.

You need

4 or 5 different types of feathers (turkey, dove, chicken)
old feather pillow
feather duster
feathers used as decorations (on clothing, in a picture or from a flower arrangement)
feather that could be used as a quill (substitute black tempera paint for ink)
blunt-ended arrow with feathers
chart paper and marker

Questions you might ask

▲ For what other purpose can a feather be used?
▲ What kind of feather do you like the best?

Interest area
Group time

Science & math principle
Teaches children about the relationship between people and birds (interdependence).

Science & math skill
Develops communication skills.

Science & math attitude
Encourages children's curiosity.

▼

You need

a glass loaf (bread) pan
large bag of birdseed
calendar
ruler
tape

Measuring the Feeding Rate of Birds

S	M	T	W	TH	F	S
		1	2	3	4	5
6	7	8	9 2"	10 1 3/4"	11 TOM 1 1/2"	12
13	14 Mark 1"	15 Hector sue 3/4"	16 1/2"	17 1/4"	18	19
20	21 0	22	23 Kevin	24	25	26
27	28	29	30			

Activity

▲ Discuss feeding birds with the children. Talk about how birds need more food in the winter to stay warm than they do in spring or summer, and how birds keep insect populations in control.

Note: Plan to do this activity two times a year, once during cold weather and once during warm weather.

▲ With the children, tape the ruler to the inside of the bread pan so the numbers are clearly visible through the glass.

▲ With the children, fill the pan with birdseed. On the day you begin the activity, set the pan outside and write on the large calendar the level of the birdseed in inches or centimeters.

▲ On the same calendar, ask each child to write her name (or write it for her) on the day in the future that she thinks all of the birdseed will be gone.

▲ Find a covered area (to avoid rain filling the bread pan) close to the classroom and out of the mainstream of traffic.

▲ Each day, have the children check the glass pan and record the level of the birdseed on the calendar.

▲ When the birdseed is gone, check the calendar to see whose prediction was the closest to being correct (save the calendar to compare with the results you will get when doing this activity again in the spring).

Note: If you have been feeding the birds all year, the birdseed will be eaten quickly. If not, it will take a while for the birds to find the seed, and after they find it you may not notice a significant decrease in the amount of seed eaten. Remember that squirrels might be responsible for some of the missing birdseed.

Questions you might ask

▲ Why do you think birds need more food in winter?

▲ Why do you think the birdseed disappeared faster after several weeks?

Extension

Compare the results from the cold- and warm-weather feedings on a bar graph.

More challenging for older children

Use the daily calendar measurements to make a graph. Record the day and the number of inches of seed remaining on the graph.

Modifications for younger children

Rather than doing any record keeping, simply fill the pan, place it outside and check it daily. Talk about why the seed is disappearing each day.

Science & math principle

Teaches children about different kinds of feathers (diversity).

Science & math skill

Encourages children to organize what they know about feathers.

Science & math attitude

Teaches children to be comfortable with science and math.

▼

Sorting Feathers

You need

8 to 10 feathers (2 or 3 of each different kind)
tray
colored tape
tall basket for the feathers

Questions you might ask

▲ How many feathers are in each group? all together?
▲ What can you tell about a bird by looking at its feathers?

Activity

▲ Make a sorting tray by dividing a tray in sections using colored tape. Create as many sections as you have different kinds of feathers.
▲ The children sort the feathers by a similar characteristic, placing them into groups on the tray.

Extension

Make a Venn diagram. Have the children group the feathers by one characteristic in one circle, one characteristic in the second circle and a characteristic common to both groups in the middle area of the Venn diagram.

More challenging for older children

Select very dissimilar feathers so the children really have to stretch to think of how they are alike and why they are placing them in the section they chose.

Modifications for younger children

Put out feathers that are easy to sort by color or length (long, short and intermediate). Reduce the number of feathers they will have to sort to six, then add one or two more.

Seriating Feathers

Activity

▲ Place the tray with the container of feathers in the Manipulatives Center.
▲ The children put the feathers in order of size from largest to smallest and then from smallest to largest on the tray.

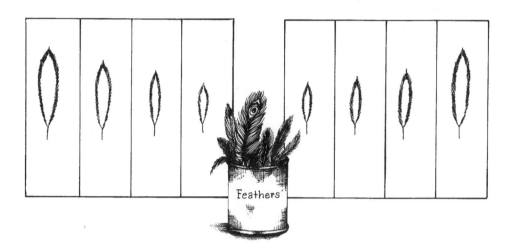

Extension

Compare the feather seriation to seriating the texture of fabric (from smooth to rough). Use fabric pieces that are relatively the same size so that the children can seriate them by feeling the texture. How does feeling the fabric compare to the way they seriated the feathers?

Note: The children are comparing their visual observations with their feeling (tactile) "observations" to seriate. Have them talk about which is more reliable or accurate and why.

More challenging for older children

Seriate the feathers by length and then by width. Did the seriation change? If so, how?

Modifications for younger children

Make a silhouette board for the children to match the feathers from largest to smallest. Give them three feathers. Have them find the biggest, the middle-sized and then the smallest. Later, give them all of the feathers to see if they can seriate them. Many will be able to do it.

Questions you might ask

▲ What kind of bird has feathers like these?
▲ What do you like about the feathers? dislike?

Interest area
Outdoors

Science & math principle

Teaches children about the many types of birds (population).

Science & math skill

Encourages communication skills.

Science & math attitude

Teaches children respect for birds.

▼

You need

index cards, 1 per child
markers, 1 per child

A Bird Walk

Activity

▲ Make tallying cards by drawing an eye in the upper left-hand corner of one side of the index card and an ear in the opposite corner (see illustration).

▲ Take a walk around your neighborhood looking for birds. If the children hear a bird, they put a mark on the ear side of the card. If they see a bird, they put a mark on the eye side of the card.

▲ When you return to the classroom, the children count up how many birds they heard and how many they saw.

▲ Talk about the number of birds seen and the number of birds heard. Discuss which has the greatest number of tallies and why.

See	Hear
|||||	||

Questions you might ask

▲ Why do you think you saw more birds than you heard? (or heard more than you saw?)

▲ Do you think you saw different kinds of birds or many of the same kind?

Extension

Try to identify the birds that live in your neighborhood.

More challenging for older children

Make a chart or graph of the children's individual findings.

Modifications for younger children

Take the neighborhood walk and keep one tally sheet for the class.

Bird Dinner

sunflower seeds

mixed

cracked corn

other

Which will the birds like best?			
Sunflower	mixed	cracked corn	other
Joe Hector Carl	Sarah Harvey	mary sam	Pete

Interest area
Outdoors

Science & math principle
Teaches children about the many types of birdseed (diversity).

Science & math skill
Develops observation skills.

Science & math attitude
Develops children's patience.

▼

You need

shelled sunflower seeds
mixed birdseed
cracked corn
4 aluminum pie pans
chart tablet and marker

Activity

▲ Ask the children to fill each pie pan with one type of feed.
▲ Encourage them to predict which type of feed the birds will like the best.
▲ Write their predictions on a chart.
▲ Place the pie pans in a low-traffic area outside.
▲ Check the pans daily to see which one empties first.
Note: If you have squirrels, be prepared for your research to be thrown off by their visits to the seed pan; sunflower seeds are one of their favorite foods.
▲ When the first pan is empty, check your predictions.

Questions you might ask

▲ Why did you choose _____ birdfeed?
▲ If you were a bird, which one would you like to eat?

More challenging for older children

Divide the class into four groups. Have each group keep track of one container. Have each group keep a timeline (number of days it took for the birds to eat all the food in the pan) on a piece of adding-machine tape.

Modifications for younger children

Start with two containers, then add a third and a fourth.

Extension

Get some advice from a bird expert through a local pet shop or college. Try other kinds of feeds, seeds and grains to attract different kinds of birds.

Science & math principle

Teaches children about energy as they test how the items drop to the floor.

Science & math skill

Develops observation skills.

Science & math attitude

Encourages children to be open-minded as they learn by trial and error.

▼

You need

old game board
1 large feather (a turkey feather works well)
1 book
1 poker chip (or counting bear)
book in a basket
feather in a basket

Dropping Feathers

Activity

▲ Before the children arrive, set up the game board as a screen that will hide from the rest of the children the results of an individual child's experiment with the feather and the book.
▲ Before the children do the experiment, they predict which item will fall faster, the feather or the book.
▲ They make their prediction by placing a poker chip in one of the baskets, either the basket with the feather or the one with the book.
▲ Then each child drops the book and feather at the same time (behind the game-board screen) and observes which one hits the table first.
▲ They check to see if their prediction was right.

Questions you might ask

▲ What sounds did you hear when the items were dropped?
▲ Why do you think we did this test?

Extension

Test other objects along with the feather and the book like a piece of tissue paper, a rock, a ruler, a paper clip.

More challenging for older children

Before the children perform their experiment, have them explain the reason for their prediction. Either have them record it or write it for them.

Modifications for younger children

Do this as a group activity. Have each child guess before you drop the book (using chips or counting bears involves too many steps and younger children may lose interest).

▼

Finding the Plastic Worm

Activity

▲ Before the children arrive, put the soil and the worm in the bottle. Use the hot glue gun (adults only) to secure the cap on the bottle. Put the tools on the tray.
▲ Challenge the children to find the worm in the bottle by using the tools on the tray and encourage them to discuss which tool would be the best.
▲ Have them try each of the tools to see which would be the most helpful. Talk about what birds use to find worms.

Questions you might ask

▲ Was the stethoscope helpful? Why or why not?
▲ Why do you think we sometimes say "he has an eagle eye?"

Extension

Compare bottle filled with the worm and soil to a bottle filled with birdseed and soil. Talk about which would be easier for a bird to find.

More challenging for older children

Ask the children to generate a list of other tools that might be helpful in finding worms.

Modifications for younger children

Let them explore the bottle without tools (until the newness wears off) and then introduce one or two of the tools.

Interest area
Science and math

Science & math principle
Teaches children about using a model to represent the real thing.

Science & math skill
Develops communication skills.

Science & math attitude
Develops respect for tools.

▼

You need

2-liter soda bottle
top soil to fill ¾ of bottle
long rubber fishing worm
hot glue gun
magnifying glass
magnet
prism
stethoscope
tray

Science & math principle

Teaches children about the properties of a bird's nest.

Science & math skill

Develops observation skills.

Science & math attitude

Teaches respect for animals' homes.

▼

You need

a bird's nest
tweezers
disposable gloves
magnifying glass
tray
chart tablet and markers

Questions you might ask

▲ Where do you think the birds found the materials?
▲ How do you think they got them back to their nests?
▲ How does the nest stay together?

Examining a Bird's Nest

Activity

▲ Place the bird's nest (found on the ground, never in a tree) and the materials listed above on a tray.
▲ The children put on the disposable gloves and use the tweezers to disassemble the nest.
▲ They examine the different materials from which the nest is made with a magnifying glass when they have separated the nest into its components.
▲ Encourage them to put the objects that they think are twigs in one pile; string, yarn or thread in another pile; leaves and grass in another; and everything else in the last.
▲ Make a chart listing the different items they have found in the bird's nest.
Note: Make sure the children wash their hands thoroughly after examining the bird's nest even though they have worn gloves.

Extension

Gather together the same materials found in the nests and encourage the children to build a nest like the bird's.

More challenging for older children

If it is possible, ask parents to bring in nests that have fallen on the ground. Divide the children into small groups to work with the different nests. Have each group make their own list of materials that birds use to make nests and have them compare it to the other groups' lists.

Modifications for younger children

Do this in a small group setting and supervise closely. Make sure the children wear the gloves and that they wash their hands thoroughly when they have finished exploring the nest.

Soaking a Bird's Nest in Water

Interest area
Science and math

Science & math principle
Teaches children about the sprouting cycle.

Science & math skill
Develops observation skills.

Science & math attitude
Develops children's patience.

▼

Activity

▲ Place the bird's nest in the tub of water. The water level should extend about half way up the side of the nest.
▲ Ask the children to predict what they think will happen as a result of the soaking. Record their predictions on paper.
▲ Leave the nest in the tub for several days and maintain the same water level.
▲ Each day observe the nest to see what has changed. If the nest is not too old, you will see sprouting seeds. This may lead to a discussion of why this happened.
▲ Encourage the children to use the magnifying glass to examine the seeds as they grow.

You need

fresh bird's nest
small tub
water
magnifying glass

Extension

Find a vacant wasp nest or a dirt dauber's nest. (Dirt daubers are also called mud daubers. They make nests in mud cells and fill them with spiders and insects so their babies will have food to eat. They are in the wasp family.) Soak it like the bird's nest and compare them.

More challenging for older children

Ask the children to represent the steps of the sprouting process by drawing a series of pictures.

Modifications for younger children

Place the tub of water with the nest in it where the children can see but not reach it (inside an empty aquarium would be a good place).

Questions you might ask

▲ How do you think the seeds got into the nest?
▲ Where else do you think the birds might drop their seeds?

Note: Make sure the nest was found on the ground and not taken from a tree. The nest is too old if it has lost the soft parts that typically layer the bottom.

▼

Orange Cup Bird Feeder

Interest area
Science and math

Science & math principle
Teaches children about change as the birds eat from the orange feeder.

Science & math skill
Develops observation skills.

Science & math attitude
Develops children's patience.

▼

You need

half an orange (1 per class or 1 per child)
2 pieces of string, each 36" (1 m) long
large, plastic needle with blunt end
knitting needle
large cork

Questions you might ask

▲ What kind of birds like the orange?
▲ How many birds do you think it will feed?

Extension

Hang other fruit from the tree and add peanut butter.

Activity

▲ Thread the needle with one piece of the string.
▲ Insert it through the orange half about ½" (13 mm) from the top of the sliced end.
▲ Pull the string until it is even on both sides.
▲ Rotate the orange ninety degrees and insert the second piece of string the same way. Pull the string until it is even on both sides. You should have four string ends, extending from the sides of the orange half, about ninety degrees apart.
▲ Tie the four ends together so you can hang the orange half from a tree limb.
▲ Insert a knitting needle through the orange between any two of the strings, at the midpoint of the orange half. The knitting needle provides a perch for the birds.
▲ Place a cork over pointed end of the knitting needle. Hang the orange in a tree. Many birds will eat the fruit and enjoy the feast.

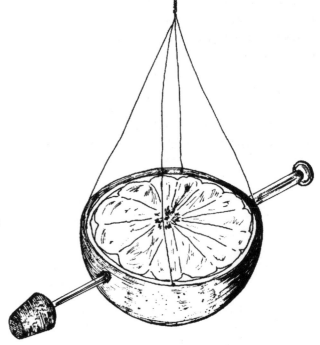

▲ Check the orange every few days with the children to observe the changes that have occurred.

More challenging for older children

Place the orange where it is visible from the classroom. Have binoculars nearby. If the children see a bird eating the orange, they report it to the class. Have them draw a picture of the birds they see eating the orange.

Modifications for younger children

Divide the class into small groups and have each group make the orange bird feeder. To attract the birds more quickly, place birdseed on the exposed fruit (top) of the orange. Otherwise it may take the birds time to find the orange, and it is hard for younger children to wait.

Recipe for the Birds

Activity

▲ Gather all of the children together to watch and to participate, depending on their ability level. Crush dog biscuits and peanuts with the mallet on the chopping board. Set aside.

▲ Pour raisins and wheat germ into the large mixing bowl. Add the crushed dog biscuits and peanuts. Add the sunflower seeds. Mix well.

▲ Heat corn oil and honey on a stove or a single heating element.

▲ Pour the heated corn oil and honey mixture into the dry mixture in the large mixing bowl and stir well.

▲ Pour the combined ingredients onto a baking tray and spread evenly. Bake in an oven at 375°F (190° C) for 10 minutes. Let it cool.

▲ Place a small amount in a small tray or bird feeder and put it outside for a bird snack. Replenish the tray when the birds have eaten it all.

▲ Predict how long it will take for the birds to find the food and to finish the food. Write the children's predictions on the chart tablet. Keep track of the birds and compare the predictions to the actual events.

▲ When you replenish the supply, ask the children if it will take the same amount of time for the birds to finish it. Record predictions and compare with results.

Questions you might ask

▲ Which ingredient did we use for the birds that you would like to taste?

▲ Why do you think the birds might like this recipe?

Extension

Find out about other recipes from a bird expert. See which ones the birds like the best.

More challenging for older children

Ask the children to draw their own rebuses, retelling with pictures how to make this recipe for the birds.

Modifications for younger children

Let the children mix the recipe with their hands. Young children really enjoy tactile, sensory experiences. Note: There is nothing in this recipe that would be injurious to children, so do not worry if a child sneaks a bite!

Interest area
Science and math

Science & math principle
Teaches children how birds and humans need each other (interdependence).

Science & math skill
Develops observation skills.

Science & math attitude
Teaches comfort with science and math.

You need

1 cup (250 ml) dog biscuits
1 cup (250 ml) unsalted peanuts
1 cup (250 ml) raisins
1 cup (250 ml) wheat germ
1 cup (250 ml) shelled sunflower seeds
½ cup (125 ml) honey
½ cup (125 ml) corn oil
chopping board & mallet
large mixing bowl and spoon
2-quart (2 L) pot
stove or hot plate
measuring cups
baking tray
oven
small tray or bird feeder
chart tablet and markers

Homemade Bird Food

pot

Wheat germ

dog biscuits

sunflower seeds

CORN OIL

Honey

unsalted peanuts   raisins

mixing bowl

Spoon

mallet

½ cup   1 cup   tray

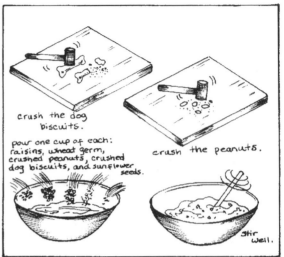

crush the dog biscuits.

pour one cup of each: raisins, wheat germ, crushed peanuts, crushed dog biscuits, and sunflower seeds.

crush the peanuts.

stir well.

Heat ½ cup of corn oil and ½ cup of honey.

pour warm mixture over other mixture and stir well.

pour mixture on the tray. spread evenly.

Bake in oven at 375° for 10 minutes.

Let the mixture cool.

Feed the birds.

Birdseed Frozen in Winter

Activity

Note: The point of this activity is to make children aware that birds have a much more difficult time finding food in winter. We want them to appreciate this fact.

▲ Mix ½ cup (125 ml) of soil, one cup (250 ml) of birdseed and ¼ cup (60 ml) of water in one of the cake pans. Mix well. Place the cake pan in the freezer overnight.

▲ Before the children arrive, place the following on a tray: one cup (250 ml) of plain birdseed in a cake pan, the frozen birdseed mixture in the other cake pan, the two tweezers and the two containers.

▲ Two children play the role of being birds. One bird is feeding in summer and the other bird in winter. They will use the tweezers as beaks. One child will try to remove the seed from the frozen cake pan while the other will try to remove seed from the pan of loose birdseed. They both pick up the birdseed with the tweezers and put it into the butter tubs.

▲ At the end of one minute, count (or just look at) the number of seeds in each tub and compare the results.

State the Problem or Question

Which is easier for the birds to pick up, the frozen birdseed or the loose birdseed?

State the Hypothesis

We believe it is easier for the birds to pick up _____. (Let the children make individual or class predictions.)

Method of Research

We will have two cake pans and two children playing the role of birds trying to feed on the seed. The two children will use pointed-end tweezers as beaks. In one pan we will freeze birdseed and soil; in the other pan we will have loose birdseed. The children will use the tweezers to pick up the birdseed in each pan. We will compare our results after we count how many seeds the two children were able to pick up in one minute.

Checking the Hypothesis

We will look at our predictions and see if we were correct.

Results

We found that it was easier to pick up birdseed that was _____.

Interest area
Science and math

Science & math principle
Teaches children how birds and humans need each other (interdependence).

Science & math skill
Develops observation skills.

Science & math attitude
Encourages children's curiosity (a desire for knowledge).

You need

½ cup (125 ml) of soil
2 cups (500 ml) of birdseed
2 round aluminum cake pans
2 tweezers with pointed ends
2 small containers (like a butter tub)
large tray
water
freezer

Questions you might ask

▲ Why is it easier to pick up birdseed that is loose?
▲ How could you help birds at your home? Do birds need more help in winter time? Why?

Extension

"Reinvent" the bird beak by trying other tools to see if they are more effective than pointed-end tweezers in gathering birdseed in the winter.

More challenging for older children

Encourage the children to make a list of other ways birds can be helped in the winter (types of food or special shelters).

Modifications for younger children

Tweezers are hard for them to manipulate. Let them use their fingers to pick up the seeds. While this is not as accurate, it will accomplish the goal of encouraging them to think about the needs of birds and other animals, and how to meet their needs in the most direct way (using science to know how best to help them feed).

The X-ray Box

Activity

▲ Cut out the top and the bottom of the box.

▲ Staple the strips of fabric across the top of the box, back to front (see illustration).

▲ When the children arrive, have them help you make an "x-ray machine." First, ask them to examine the skeleton or skeleton picture and count the ribs.

▲ Next, ask them to decide which part of the skeleton would show if one of the children hung the box from his shoulders.

▲ Encourage them to paint the box front like the skeleton they imagine they would see.

▲ Use this x-ray machine in the doctor's office of the Home Center. The children slip into the box and pretend to have an x-ray made.

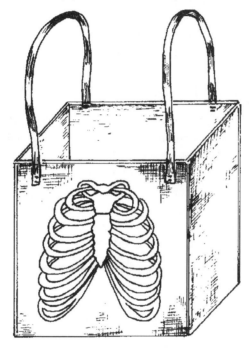

Questions you might ask

▲ How is this x-ray machine different from a real one?

▲ Can you think of another way to make a picture of the inside of a body?

Extension

Create the props needed for a doctor's office.

More challenging for older children

Divide the class into groups of four or five. Give each group a box, a skeleton picture and paints to draw skeletons. Compare the group's results after they have made their x-ray boxes.

Modifications for younger children

Give each child a smock. Children paint their skeletons on the smock (please note that younger children will not be able to paint a skeleton in any recognizable manner).

Interest area
Home center

Science & math principle
Teaches children about models by making a representation of a bone.

Science & math skill
Encourages children's communication skills when they make boxes.

Science & math attitude
Teaches children to respect craftsmanship.

▼

You need

cardboard box, 12" x 16" x 21" (30 cm x 40 cm x 53 cm)
sharp cutting tool
white tempera paint
several paintbrushes
2 strips of wide fabric or ribbon, 12" (30 cm) long
stapler
skeleton or picture of a skeleton

Interest area
Manipulatives

Science & math principle
Teaches children about the properties of bones.

Science & math skill
Encourages children's observation skills.

Science & math attitude
Develops children's natural curiosity about bones.

▼

You need
large x-ray film
black permanent marker
scissors
tray

A Puzzling X-ray

Activity

▲ Before the children arrive, cover any name on the x-ray with black permanent marker and cut the film into three, five, seven or more pieces (depending on how hard you want the puzzle to be).
▲ Place the pieces on a tray and place the tray in the Manipulatives Center for the children to examine and ask questions about (not only of you, but of their classmates, from whom they will learn a great deal).
▲ Have the children assemble the puzzle.

Questions you might ask

▲ How are the x-ray pictures of bones like the bones in your body?
▲ Why do our bodies have bones?

Extension

Cut apart several x-ray films and mix them up. Have the children find and assemble the different x-ray pieces for the different puzzles.

More challenging for older children

Cut the x-ray film into twenty to thirty pieces. Obtain a second x-ray film of the same body part. The children compare the second film to the first, finding similarities and differences.

Modifications for younger children

After you cut the film into puzzle pieces, trace the outline of each piece of the puzzle, in assembled form, on a large sheet of white paper. This gives younger children a base (and reference) upon which to work the puzzle.

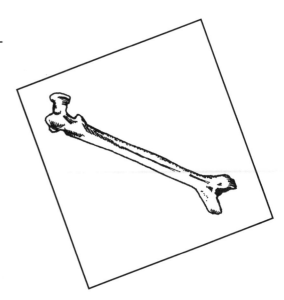

Two Known Bones

Activity

Note: The bones in each pair should be a close match.
▲ Place the bones in the basket and the basket on the tray.
▲ Put the tray in the Manipulatives Center.
▲ The children find the bones that match and pair them.

Questions you might ask

▲ What do you think this bone (point to one) does for the body?
▲ Pretend you are a bone. What would you look like? How big would you be? What job would you do?

Extension

Bury the matching bones in a tub of sand. Have the children feel the bones in the sand to find the matching pairs (no looking!). Have the children remove and verify only their sense-of-touch "observations."

More challenging for older children

Use bones that are very similar to make pairing more difficult.

Modifications for younger children

Choose matching bone pairs that are grossly different. (Have pairs where each one of the pair is very similar, but where the pairs themselves are grossly different from each other.)

Interest area
Manipulatives

Science & math principle
Teaches children about the patterns in bones.

Science & math skill
Encourages children to compare the bones.

Science & math attitude
Develops children's curiosity about bones and their characteristics.

You need

turkey or chicken bone pairs
tray
basket

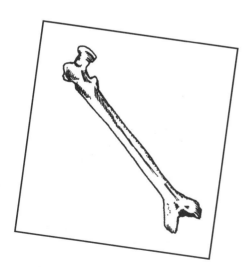

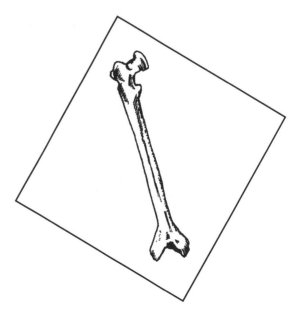

Interest area
Music and movement

Science & math principle
Teaches children to create a pattern.

Science & math skill
Develops children's observation skills.

Science & math attitude
Teaches children to value and respect craftsmanship.

▼

You need

6 or 7 different-size bones
36" (1 m) length of string
tape player and music tape
2 nails

Musical Bones

Activity

▲ Tie together all but one of the bones using the string.
▲ Tie the string of bones to two nails so it hangs across a corner of the room, giving it room to swing freely.
▲ Put on some music.
▲ The children use the extra bone to tap the other bones to music.

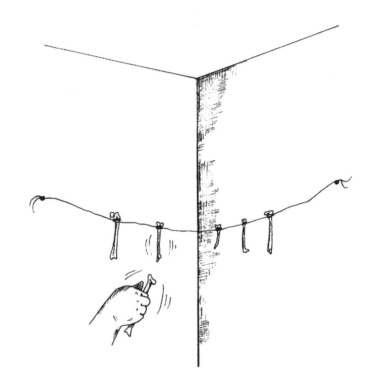

Questions you might ask

▲ Do all the bones sound the same when they are tapped?
▲ What other things could you tap to make music?

Extension

Hang bones and other things like spoons or twigs for the children to tap. Have the children compare the sounds of striking different objects.

More challenging for older children

Encourage them to tap out the sound of a song like "Row, Row, Row Your Boat" without the music.

Modifications for younger children

It might be easier to strike bones with a wooden spoon instead of another bone. You might want to set this up so the bones do not have a big "swing way."

Bone Graphs

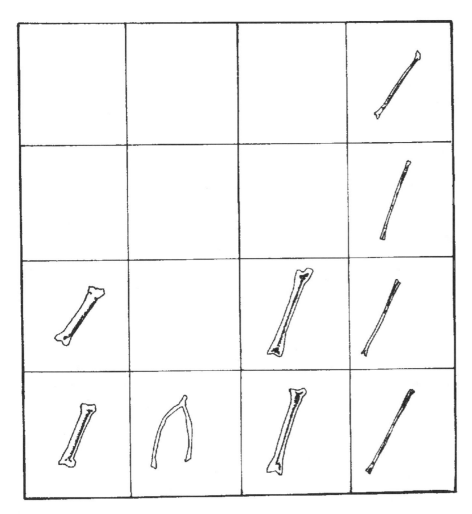

Science & math principle

Teaches children about the diversity of bones, how there are many types of bones.

Science & math skill

Encourages children to communicate the results in a graph.

Science & math attitude

Develops children's curiosity about bones.

▼

You need

several different bones (chicken, turkey, beef or lamb bones can be obtained from your butcher or at the grocery store)
window-shade graphing grid from Rock Graphing, page 197

Activity

▲ The children place the bones on the grid according to bone characteristics, such as straight or curved, long or short, etc.

More challenging for older children

Choose bones that are very similar so the older children will have to look closely and think carefully about the characteristics by which the bones are to be graphed.

Modifications for younger children

Choose bones that are very different. It will be easier for them to see similarities.

Questions you might ask

▲ Which bones are your favorites?
▲ How are all of the bones alike? different?

Extension

If possible, find bones from different animals and graph them.

The Bone Bath

Interest area
Science and math

Science & math principle
Teaches children about how bones can change.

Science & math skill
Encourages children to be careful observers.

Science & math attitude
Develops children's patience as they wait for the results.

▼

You need

2 small bones
vinegar
water
2 jars with lids
chart tablet and marker

Activity

▲ Put one bone in each jar.
▲ Fill one jar with vinegar, the other with water.
▲ Put on the lids of both jars.
▲ Ask the children to predict what will happen to the bones if you give them a "bath" for two weeks, putting one bone in vinegar and the other in water.
▲ Write down their predictions.
▲ Place the bones in the jars.
▲ Check them every day to see what has changed.
▲ After two weeks remove the bones and have the children examine them. What are the differences?
▲ Check their predictions and have them guess why the bone in vinegar became soft and pliable while the bone in water did not.

Note: The vinegar, an acid, changes the calcium phosphate in the bone (which makes bones hard) into calcium acetate. Calcium acetate is water soluble, so gradually it dissolves in the water. This leaves the bone soft and pliable, like rubber.

Questions you might ask

▲ What happened to the bone bathed in water? in vinegar?
▲ What would happen if they were left in the jars for two or three months?

Extension

Allow the bones to remain in the jars for three or four months. Observe what happens.

More challenging for older children

Have the children use a number line and record by day their observations of the changes to the bones during the two weeks.

Modifications for younger children

Have a wet wash cloth handy when they feel the softened, pliable bone. Their hands will smell like salad! Do not try to record their predictions, just talk about them and make your observations each day.

Inside Saw-bone-ing

Activity

▲ Talk to the children about what bones are. What are bones for? Why are they important? What happens when we break a bone? Talk about how bones grow.

▲ Saw through several bones for the children to examine. This is simple to do with a hand saw (adult only) when the bones are dry.

▲ Look inside and outside the bone shaft. Ask them what they see and how it feels when they touch it.

▲ Place the bones on a tray with two bones that have not been sawed in half.

▲ The children use the magnifying glass to make comparisons.

Interest area
Science and math

Science & math principle
Teaches children about the properties of bones.

Science & math skill
Helps children realize how important observation skills are.

Science & math attitude
Teaches children to be comfortable with science and math.

▼

Questions you might ask

▲ What does the inside of the bone look like?

▲ Are all the bones different on the inside or are they alike?

Extension

Examine a fresh bone and an old sundried one. Old bones have empty centers. The marrow has dried and all you find is an empty shaft where the marrow had been.

More challenging for older children

Ask the children to draw the inside and the outside of a bone.

Modifications for younger children

Keep the saw out of reach of the children. Make sure younger children understand that if they break a bone, the bone grows back together again. Let the children hold bones while you are talking to them. Be sure to have enough bones for everyone!

You need

several different bones (chicken, turkey, beef or lamb bones can be obtained from your butcher or at the grocery store)
small saw (for adult use only)
tray
magnifying glass

Examining Owl Pellets

Science & math principle

Teaches children about the properties of owl pellets.

Science & math skill

Develops children's observation skills.

Science & math attitude

Encourages children's curiosity.

▼

You need

owl pellets
rubber gloves
tweezers
magnifying glass
tray
white paper

Questions you might ask

▲ Which bones are your favorites?
▲ How are all of the bones alike? different?

Extension

If possible, find bones from different animals and graph them.

Activity

▲ Place the white paper on the tray along with an owl pellet, tweezers and rubber gloves.
▲ The children examine the owl pellet in its whole form.
▲ Talk about what they see.
▲ One child at a time puts on the gloves and uses the tweezers to separate the parts of the pellet and spread them on the white paper.
▲ Help the children find animal bones, feathers, fur, etc., and to examine them separately.
▲ Place the tray (with all of the pellet parts displayed), the gloves, the magnifying glass and the tweezers in the Science Center for the children to examine freely.

Note: An owl pellet is a regurgitated residue of fur, feathers, bones and teeth of small animals and birds on which owls feed. Owls swallow their prey whole. The fleshy parts are digested in a few hours by the strong acids in the owl's stomach. The fur, feathers, teeth and bones cannot be digested, so they accrete into a pellet which periodically is regurgitated. It dries to a gray-brown color. Sometimes pellets can be found on old barn floors or beneath the trees in which owls roost. Owl pellets also can be ordered from science catalogs, which is the simplest way to obtain them. When they are obtained by catalog, a description list of the contents is sent along with the pellet(s) to help identify what is being seen. When the children sort through the bones, they will likely find skulls, jaws, ribs and backbones of shrews, moles, rats, sparrows, weasels or voles.

More challenging for older children

Ask the children to predict what they will find in the owl pellet. Have them describe what they actually found by drawing pictures of their findings.

Modifications for younger children

Do this activity in groups of three or four so each child can get a good view of the examination. Have a pair of gloves for each child so that they can immediately touch the pellet and the parts. Most of the children will not be able to manage the tweezers, so let them use their gloved fingers.

Bone Sorting Chart

	RODENTS	SHREWS	MOLES	BIRDS
SKULLS				
JAWS	loose teeth			
SHOULDER BLADES				
FRONT LEGS				
HIPS				
HIND LEGS				
ASSORTED RIBS				
ASSORTED VERTEBRAE				

bones

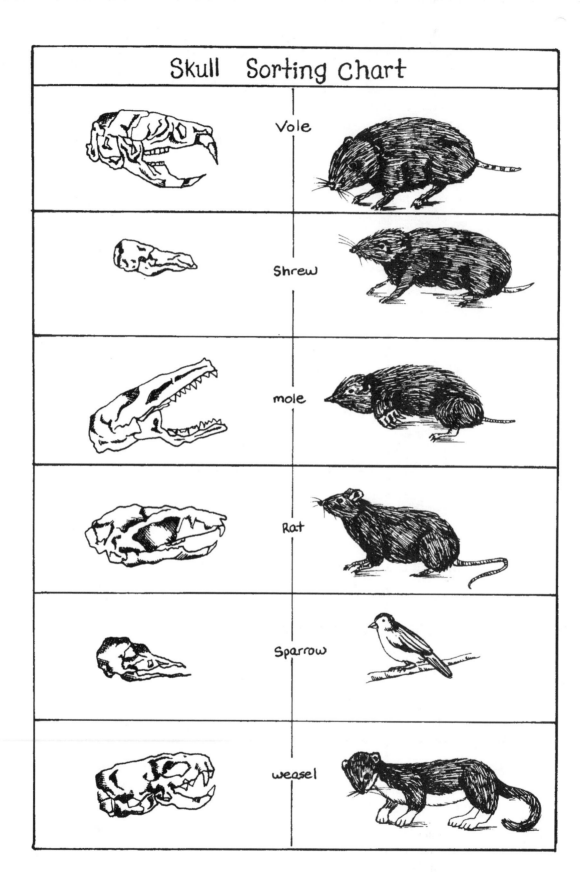

Skull Sorting Chart

	Vole	
	Shrew	
	mole	
	Rat	
	Sparrow	
	weasel	

Turkey Bones & Ant Mounds

Activity

▲ Find an ant mound outside that is easy to observe and will not be disturbed by other children.
▲ Place the carcass about 6 inches (15 cm) away.
▲ Cover the carcass with an old aquarium (upside down).
▲ The children observe the ants cleaning the bones over time.
▲ Mark down the changes that occur daily on a calendar and keep a number line or a daily log.
Note: Seal the aquarium edges with a little soil to keep the odor from attracting other animals to the turkey carcass.

State the Problem or Question
What will happen to the turkey carcass when it is placed close to an ant mound?

State the Hypothesis
We believe _____ will happen to the carcass. Have each child make a prediction and write all the predictions on a chart. Post the chart in the classroom.

Method of Research
We will place the carcass underneath an aquarium and put it next to an ant mound. We will observe the changes that take place each day. We will leave the carcass underneath the aquarium until all of the turkey meat has been removed from the turkey bones.

Checking the Hypothesis
After the turkey bones are cleaned, we will check our predictions to see if they were correct.

Results
We found _____, _____ and _____ happened when we left the turkey carcass under the aquarium and the ants cleaned the bones.

Extension

Do the same activity with two turkey carcasses and two aquariums. Place one carcass as discussed above; place the other one about 5 feet (1.5 meters) away from the first aquarium. Compare the results.

More challenging for older children

Ask the children to write a journal of their observations of the changes that take place as the ants clean the turkey meat from the bones.

Modifications for younger children

Since the changes that take place are gradual, observe weekly rather than daily. Rather than keeping a calendar or a number line, just talk about what is happening to the turkey bones.

Interest area
Science and math

Science & math principle
Teaches children about how bones can change.

Science & math skill
Develops children's observation skills.

Science & math attitude
Develops children's patience as they wait for the results.

You need

turkey carcass with very small bits of meat left on the bones
old aquarium
ant mound outside
calendar or piece of paper
chart

Questions you might ask

▲ Why do you think the ants liked the carcass?
▲ Why do you think the bones are not laying on the ground?

Science & math principle

Teaches children about cause and effect.

Science & math skill

Encourages children to sharpen their observation skills.

Science & math attitude

Develops children's curiosity.

▼

You need

dried corn cobs (use those from Husks, Shucks and Corn on the Cob, page 84)
tray partially filled with tempera paint
paper
newspaper

Corn Cob Applications

Activity

▲ Place the tempera paint tray on a newspaper-covered table.
▲ Encourage the children to grasp the corn cob, dip one end into the tempera paint and print on the paper to make a design.
▲ As the children work, talk about the number of prints they can make with each dip into the paint (without redipping) and about where they are making their print in relationship to the prints they made before.
▲ Ask them about the size of the prints they are making and about the frequency on the page. For example, are the prints side-by-side, overlapping one another, separated a little, separated a lot?

Questions you might ask

▲ How are your prints alike? different?
▲ What is the difference between painting and printing?
▲ How else could you make a print? a design of prints?

Extension

Print with both ends of the corn cob. Roll the cob in the paint and then roll it on the paper. Combine methods to make a roll-print design.

More challenging for older children

Ask each child to write a story about her print or design.

Modifications for younger children

Use large sheets of paper. Expect that they will paint rather than print with the corn cob, and have lots of wet paper towels handy to clean their hands.

Corn-Shuck Tie-Up

Activity

▲ The day before doing this activity, place the corn shucks in the dishpan and cover them with water, leaving them to soak overnight. This softens the shucks and makes them pliable.

▲ Place the shucks in one basket lined with paper towels.

▲ Place the twist ties, rubber bands, scissors and markers in the second basket.

▲ The children twist and fold the shucks and apply a rubber band or a twist tie at any point on the shucks to hold them together.

▲ The children might want to cut their shucks or draw on them.

▲ Set their constructions aside to dry.

Note: Some children might discover they can make a corn shuck doll by using more than one rubber band or tie. As the children work with the corn shucks, they are exploring the quantities and qualities of things and learning about roundness, flatness, thickness and texture. As they apply the rubber band or twist tie, they are learning also about length and width. They are learning to improvise when they have to make adjustments to complete an idea they have in their heads.

 Cornshuck Construction

Get and rubber bands

or twist ties

Make a design

Wrap with rubber bands or twist ties

Extension

Explore working with the husk dry and the husk wet. Compare the difference in the materials and in the results of their work.

More challenging for older children

Encourage the children to work in small groups to see what they can make by working together and combining raw materials to make one thing. Let them brainstorm.

Modifications for younger children

This is a difficult activity for small, developing hands. Do not use rubber bands. Get the long twist ties and allow them to help each other to accomplish their tasks.

Interest area
Art

Science & math principle
Teaches about the properties of the corn shucks.

Science & math skill
Encourages the children to learn about corn shucks by observation.

Science & math attitude
Develops the children's confidence as they work with the corn shucks.

▼

You need

corn shucks (you can purchase them in bags at the grocery store)
water in a dishpan
2 baskets
paper towels
rubber bands
twist ties
scissors
markers

Questions you might ask

▲ How does the corn shuck feel? Have you ever felt anything like it before?

▲ What was the easiest part of the activity? the hardest?

▼

Cornbread Cupcakes

Interest area
Cooking and snack

Science & math principle
Teaches children about changes that take place when cooking cornbread.

Science & math skill
Encourages children to compare their mixture to the finished cornbread.

Science & math attitude
Encourages children to respect the tools used to make cornbread.

You need

ready-mix cornbread, 2 level tablespoons (30 ml) per child
tablespoon measure
wax-coated paper cups
electric skillet
milk
spoon
craft sticks (1 per child)
small container of sugar

Activity

▲ Use the rebus (picture drawing) to explain how to cook the cornbread in individual cups.
▲ Set up the activity in a station format as follows:

Station 1
Put 2 level tablespoons (30 ml) of mix in a cup. Have a plastic knife handy so the children will be able to level their measure.

Station 2
Add a "pinch" of sugar (estimation in action).

Station 3
Add 1 tablespoon (15 ml) of milk. Stir with a craft stick. If it seems too dry, add a little more milk. You want the mix thick, but smooth.

Station 4
Wait until five more children have mixed cornbread in cups, then cook six cups at a time in the electric skillet.

▲ Preheat the skillet to 400° F (200° C); put the lid on the skillet and cook for 10 to 12 minutes. The cornbread in the bottom of the cup obviously will be the first to burn, so check the bottom of each cup at 10 minutes.
▲ Set the cornbread aside for 10 to 15 minutes to cool. When cool, the children peel off the paper and eat the cornbread.

Note: Since each skillet is different, you may want to test the baking time of your skillet at home before attempting this activity at school with the children. When it is done, the cornbread will spring back when you press the bread surface with your finger.

Questions you might ask

▲ Why do you think people call it cornbread?
▲ How is this cornbread like other cornbread, or bread, you have eaten?

Extension

Many cake mixes can be cooked in a cup. When the recipe calls for an egg, just add a bit more water or milk and omit the egg or beat the egg thoroughly and have the children add an eyedropper full of egg to their cup.

More challenging for older children

Let the children organize the stations and make their own picture directions for this activity.

Modifications for younger children

Do the activity in small groups and omit the work station format. You will have to help the group perform certain tasks. Put a red masking-tape square around the skillet to create a "no touch" zone.

you need:

corn Bread mix

Milk

Sugar

Station 1

corn Bread mix

put 2 tablespoons of mix in the cup

Station 2

Sugar

put in a pinch of sugar

Station 3

Milk

put in 1 tablespoon of milk and stir

cook at 400° for ten minutes

▼

Making Corn Tortillas

Interest area
Cooking and snack

Science & math principle
Teaches children about how cornmeal changes when cooked.

Science & math skill
Encourages children to compare the cooked and uncooked mixtures.

Science & math attitude
Develops children's desire for knowledge and their curiosity.

You need

corn flour (also called masa)
hot water
measuring cup
salt
electric skillet or a comal
large bowl and spoon
rolling pin
spatula

Questions you might ask

▲ What does the corn flour feel like when it is dry? wet?
▲ Have you ever tasted anything like tortillas before?

Activity

▲ Mix together two cups of corn flour (do not use cornmeal), a pinch of salt and 1 to 1½ cups (250-375 ml) of warm water (almost hot to the touch).
▲ Mix well in a large bowl and knead with your hands.
▲ When it is the consistency of pie-crust dough, the children roll it into same-size balls, then press them flat. This recipe will make ten 6- to 8-inch (15-20 cm) tortillas or fifteen 4-inch (10 cm) tortillas.
▲ Sprinkle corn flour on the work surface and on the rolling pin. The children spread out the flattened balls even further with the rolling pin.
▲ Cook the tortillas on one side in the hot skillet or comal until they are golden brown. Flip them with the spatula and cook them until they are brown on the other side.
Note: A comal is a flat cast-iron skillet with no sides. You can also use an electric skillet or a cast-iron skillet.

Extension

Use the corn tortillas to make tacos, nachos and tacquitos (corn tortillas rolled up with meat sauce or refried beans inside and then deep fried).

More challenging for older children

Ask the children to talk about flour tortillas. How does the taste compare to corn tortillas? Which one sticks to the roof of their mouth when they eat it? Which one folds easiest?

Modifications for younger children

The children will need help rolling the dough balls and using the rolling pin. Place a red tape square around the cooking element, making a "no touch" zone.

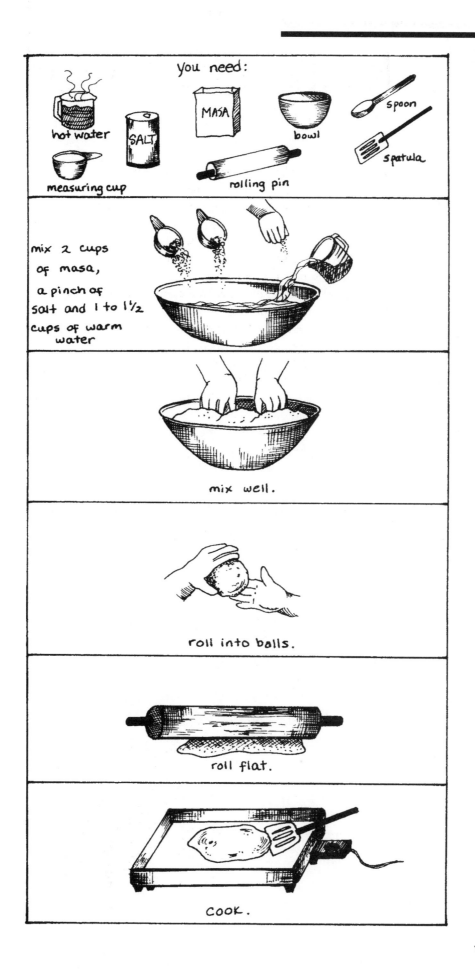

you need:

hot water · SALT · MASA · bowl · spoon · spatula · measuring cup · rolling pin

mix 2 cups of masa, a pinch of salt and 1 to 1½ cups of warm water

mix well.

roll into balls.

roll flat.

cook.

Husks, Shucks & Corn on the Cob

Interest area
Cooking and snack

Science & math principle
Teaches children about how corn changes.

Science & math skill
Encourages children to use their observation skills.

Science & math attitude
Develops their curiosity when they cook the corn.

▼

You need

1 ear of corn with husks on
several bags of frozen ears of corn (1 ear per child)
water for cooking corn
pot
salt
margarine
bamboo skewers
paper plates
stove or hot plate

Questions you might ask

▲ Have you ever seen anything like this before? Where?
▲ How did the corn change after cooking it?

Activity

▲ Talk about the ear of corn in its husk.
▲ Let the children guess what is inside the husk.
▲ Have each child peel back a layer of husk from the ear without removing it, until all the husk has been peeled back and the ear is exposed.
Note: Save this ear that has been partially shucked for Dried to a Pucker, page 91. Fold the husk back over the ear, put it into a resealable plastic bag and store in the refrigerator.
▲ Encourage the children to feel the corn and to talk about what it looks like.
▲ Talk about the cornsilk and the husk.
▲ Set the ear and husk aside and show them the small ears of frozen corn.
▲ Explain that you are going to cook them in boiling water, according to the directions on the package.
▲ Boil the corn.
▲ While the ears are hot, push bamboo skewers through each ear. Do this yourself because the skewers are very sharp.
▲ Snip off the pointed ends of the skewers, place corn on paper plates and pass out, one ear for each child.
▲ Put out margarine and salt for those children who would like them.
Note: Save the cobs that the children have finished for Corn Cob Applications, page 78. With the skewers in place, wash the cobs and pick off any whole kernels left on the cob with your fingers (do not scrap with a knife). Stand them up by pushing one end of each skewer into a large block of Styrofoam, allowing for space between cobs. Let them dry for four or five days in a sunny spot in the classroom.

Extension

Scrape off the raw corn from an ear and cook it in boiling water. Compare the taste of it to corn on the cob.

More challenging for older children

Ask the children to make a list of why they think we call this form of corn an "ear" of corn and why we call it "corn on the cob."

Modifications for younger children

Supervise this activity well. Even with their ends cut off, the skewers can be dangerous. You might prefer to stick forks into each end of the corn. Forks are easier for the children to hold. Cut the frozen ears of corn in half if your children are not big eaters. If some do not want to try the corn, that is okay.

Cornucopia

Activity

▲ Make a graph (see illustration) on the white paper.
▲ Cut out a kernel of corn for each child from the yellow construction paper and write each child's name on it.
▲ Before the children arrive, place a little bit of each corn product on a paper plate for each child.
▲ When the children arrive, talk about food products made from corn.
▲ Pass out the corn products. Encourage the children to taste each one.
▲ When they have finished, discuss the taste of each product.
▲ Ask the children to use a glue stick to place their paper corn kernel in the column of the corn product graph that indicates what they liked the best.

		Tim		
	Ari	Amed		
	Tom	Steve		Tad
Sarah	Joe	Simone	Mai	Tina
cornbread	cornflakes	cornchips	canned corn	popped corn

Extension

Put other corn products (such as a fresh ear of corn in its husk, dried Indian corn, popcorn kernels, cornmeal and a cornstalk) in the Science Center for the children to examine.

More challenging for older children

Ask the children to bring in their family's favorite corn dishes and let everyone have a taste.

Modifications for younger children

Put out just two corn products to taste. Let them choose which they liked the best. Add more food products later. Let them tell you their favorite each time.

Interest area
Group time

Science & math principle

Teaches children about the properties of things made from corn.

Science & math skill

Encourages children to communicate their knowledge of corn and corn products.

Science & math attitude

Encourages a desire for knowledge or information.

▼

You need

cornbread
corn flakes
corn chips
canned corn (heated)
popped corn
paper plates
large sheet of white paper
marker
glue stick
yellow construction paper

Questions you might ask

▲ Why do you think people cook corn in so many different ways?
▲ What do you like about the corn? dislike?

Interest area
Group time

Science & math principle

Teaches children about how many things can be combined with popcorn (diversity).

Science & math skill

Encourages children to communicate what they have learned.

Science & math attitude

Encourages children's curiosity about different toppings.

▼

You need

popped popcorn
melted caramel
parmesan cheese
melted butter or margarine
4 bowls
4 paper cups per child
large piece of white paper
marker
crayons

Questions you might ask

▲ How is each type of popcorn different from the others?
▲ What do you like about popcorn? dislike?

Popcorn Tops

Activity

▲ Make a graph like the one illustrated.
▲ Make popcorn and put equal amounts in each of the four bowls.
▲ Sprinkle parmesan cheese over the popcorn in the first bowl.
▲ Pour melted margarine or butter over popcorn in the second bowl.
▲ Pour melted caramel over popcorn in the third bowl.
▲ Popcorn in the fourth bowl should be left plain.
▲ Divide popcorn into small cups so every child gets a sample of each to taste.
▲ After the children have tasted all the popcorn, they decide which type they liked best.
▲ Ask the children to use a crayon to color in the box in one of the columns of the graph that reflects their choice of the best popcorn.

Extension

Try other toppings for popcorn. There is quite a selection in the grocery store.

More challenging for older children

Give each child four pieces of paper, with numbers from one to four written on them (one number on each piece). Ask them to rank their popcorn choices using the numbers, with number one being their favorite.

Modifications for younger children

Do this activity over a two-day period. Taste two popcorn types on one day, two more on the next. Because children may have difficulty coloring the box on the graph, give them paper that has been cut to look like the different kinds of popcorn. Let them make their popcorn preference by gluing the paper popcorn beside their choice.

A Row of Ears

Interest area
Manipulatives

Activity

▲ Make silhouettes of the corn ears by copying each ear individually on a copying machine or trace the outline of each ear with a pencil.
▲ Cut out the silhouettes and tape them to the tray.
▲ Put the corn in the basket.
▲ The children seriate the corn ears by matching the actual ear of corn to its silhouette.

Science & math principle

Teaches children about the many types of corn (diversity).

Science & math skill

Helps children learn about organizing their knowledge.

Science & math attitude

Helps children keep an open mind about the results.

Questions you might ask

▲ How are the ears alike? different?
▲ Where might the different ears grow?

Extension

Seriate by size, color gradation or thickness using only one kind of corn.

More challenging for older children

Do not use silhouettes. Have the children decide for themselves how they will seriate the ears.

Modifications for younger children

Use three ears of distinctly different sizes, then add more as the children's skills improve.

You need

5 or 6 ears of decorative corn (Burgundy, Speckled-Miniature, Chalqueno, Red Starburst and Pueblo Blue are some available types)
paper
photocopier or pencil
tray
basket
tape

The Jarred Corn Kernel Count

Interest area
Science and math

Science & math principle
Teaches children about groups of corn kernels (population).

Science & math skill
Encourages children to compare their estimate to the actual number.

Science & math attitude
Encourages children to be honest with the results.

You need
7 or 8 deer corn kernels
small jar (with a lid)
tray
adding-machine tape

Questions you might ask
▲ How did you decide on the number of kernels in the jar?
▲ Would it be easier to count them if the kernels were larger? smaller?

Activity
▲ Make a number line on the adding-machine tape, with the numbers 1-12.
▲ Place the corn kernels inside the jar and put on the top.
▲ Put the jar and the number line on the tray.
▲ Place the tray in the Science and Math Center.
▲ Ask the children to guess how many kernels they think are in the jar.
▲ They check their guesses by counting them using the adding-machine tape number line.
Note: Use deer corn because you can toss it outside and the birds will eat it. Deer corn is larger in size, and it is easier for the children to manipulate with their small hands (and their developing fine motor skills).

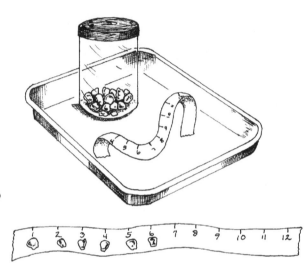

Extension
Mix different kinds of corn with the deer corn in the jar. Try popcorn kernels and dried Indian corn kernels. Have the children discuss if this makes guessing easier or more difficult.

More challenging for older children
Put in between forty and fifty kernels and ask the children to write their guesses on a piece of paper before counting them. The children practice number skills by grouping the kernels by tens.

Modifications for younger children
Put just three or four kernels in the jar. Add one or two more as their counting skills improve. Make sure the children do not put the kernels in their mouth's.

Corn Sprouts

Activity

▲ The children place the soil in the bottom of the bag and bury the kernels in the soil.

▲ Talk about what a corn kernel is (a seed that knows how to grow into a cornstalk that produces corn).

▲ Add the water and reseal the bag.

▲ Tape the bag to the inside of a classroom window through which sun shines several hours a day.

▲ On the class calendar, note the day that the bag was taped to the window. Mark off each subsequent day on the calendar on which no change is observed (after about a week or two the kernels will sprout).

▲ After the kernels have sprouted, ask one of the children to draw a picture of the sprout on the calendar.

▲ When it has grown 2 or 3 inches (5 to 8 cm) tall, ask another child to draw a new picture on the calendar.

▲ Transplant the sprout into a tire filled with dirt in the school yard. Water it every other day and then once a week when you see the sprout has established itself.

▲ Continue to keep track of the growth of the cornstalk on the calendar. Have the children measure its growth with both standard and nonstandard measurers.

Note: The growing time from kernel to sprout to young corn plant depends on the type of kernels you use and the time of year.

Questions you might ask

▲ Why do you think we put water in the bag with the kernels and the soil?

▲ How long did it take for the seed to sprout?

Extension

Plant other types of corn kernels. See if the sprouts and subsequent growth are similar to or different from those of the original corn kernel.

More challenging for older children

Each child can sprout her own corn plants and establish them in pots near a sunny window or outside.

Modifications for younger children

Do not track the growth of the sprout each day. Simply make sure the children observe and talk about it every day.

Interest area
Science and math

Science & math principle
Teaches children about the growing cycle of corn.

Science & math skill
Encourages observation skills.

Science & math attitude
Helps children develop patience as they wait for the results.

You need

1 resealable plastic bag
tablespoon measure
6 tablespoons (90 ml) of soil
2 deer corn kernels
3 tablespoons (45 ml) of water
clear tape, about 2" (5 cm) wide
tire filled with dirt
calendar

Interest area
Science and math

Corn Kernel Klasses

Science & math principle

Teaches children about the varieties of corn (diversity).

Science & math skill

Helps children organize their knowledge of corn.

Science & math attitude

Develops a desire for more information (curiosity).

▼

Activity

▲ The first part of the activity involves finding similarities and differences in the corn and discussing them.
▲ After they have examined the kernels, have children remove them from each cob with your help.
▲ For the second part of the activity, give the children a white sheet of paper and a cup of mixed corn kernels.
▲ They dump the kernels out on the white paper and sort them into five groups.
▲ Ask them to put the kernels into four groups, then into three.
▲ Ask them to explain how they grouped them.

Note: Some of the children will choose characteristics unique to their own perceptions. Also, they may forget what characteristic they used to sort the corn originally. Such events are not important. What is important is that they recognize that there are differences and similarities in things. To say it another way, we are not looking for them to sort "correctly" by predetermined characteristics.

You need

different kinds of corn on the cob (Burgundy, Speckled-Miniature, Chalqueno, Red Starburst and Pueblo Blue are some available types)
tray
sheet of white paper for each child

Questions you might ask

▲ What is one characteristic that all the kernels have?
▲ Can you think of a story about corn?

Extension

Ask older children to count the kernels from the different ears and compare the totals for each. Any significant differences? Why?

More challenging for older children

Ask the children to make a list of all of the things that could be done with colored kernels of corn.

Modifications for younger children

Use two or three ears of corn. Ask the children how they look the same and how they look different. They will especially enjoy getting the kernels off the ear. If they have difficulty, have them use both hands and wring the corn cob like they were wringing out a towel.

Dried to a Pucker

Activity

▲ The children will dry an ear of corn.

▲ Use the ear of corn from Husks, Shucks and Corn on the Cob, page 84, where the children peeled back the cornhusk but left it attached to the corn ear.

▲ Talk to the children about the water that is in corn. Recall previous activities with corn where "watery stuff" came from the corn when it was squeezed. Talk about how an ear of corn dries out (by losing water) and what it looks like when it is dried out.

▲ Run the string under the husk at the end of the ear and tie it in a loop. Hang the corn ear in a sunny window.

▲ Ask the children to predict the number of days it will take for the corn ear to dry out. Keep track of the number of days on the class calendar.

Question or statement of problem

How many days will it take to dry an ear of corn?

State the Hypothesis

We believe it will take ___ days to dry an ear of corn.

(Write predictions beside each child's name on a chart.)

Method of Research

We will tie a string to an ear of corn and we will hang it in a sunny window. We will check it each day to see the changes. We will mark it on the class calendar the day we think the corn ear is dry. We will know it is dry when the kernels are hard and there is no water left in them.

Checking the Hypothesis

We observed the corn as it dried. We noticed the days that passed by looking at our calendar and we marked them on the calendar to see if we were correct.

Results

We found it took ___ days for the ear of corn to dry.

Extension

Dry a variety of corn ears and compare the amount of time each took to dry. Talk about the reasons for the differences in drying time. Can these differences be attributed to variations in the corn, the amount of sunlight coming through the window or the weather outside? Or to all of these?

More challenging for older children

Rather than drying one ear, dry five or six. Every other week dissect one of the ears so the children can examine the kernels and the cob for moisture and record what they see.

Modifications for younger children

Puncture one kernel with your fingernail so the children can see the "watery (milk-like) stuff" that exudes from the corn kernel and that appears on the corn ear occasionally throughout the drying process.

Interest area
Science and math

Science & math principle
Teaches children about cause and effect.

Science & math skill
Encourages children to compare the ear before and after drying.

Science & math attitude
Helps children develop patience while waiting for the results.

You need

ear of corn (from Husks, Shucks and Corn on the Cob, page 84, or an ear of corn in its husk)
string
calendar
chart tablet
marker
sunny window

Questions you might ask

▲ Where do you think the water went as the corn dried?

▲ How did the corn change as it dried out?

Science & math principle

Teaches children about patterns as they sort the crayons.

Science & math skill

Teaches children to compare the crayons.

Science & math attitude

Develops children's confidence in working their own way.

▼

You need

box of 8 jumbo crayons
box of 48 crayons
large sheet of white paper
basket
tray

Encircling Crayons

Activity

▲ Draw three circles with an 8-inch (20 cm) diameter on white paper.
▲ Place the crayons in the basket.
▲ Put the basket and the sheet with circles on a tray in the Art Center.
▲ Ask the children to decide how to sort all the crayons.

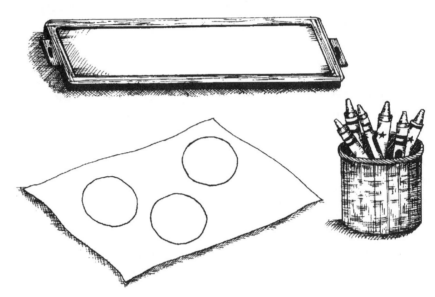

Questions you might ask

▲ How are all these crayons alike (pointing to one of the crayon-filled circles)?
▲ What other ways can you sort the crayons?

Extension

After the children have sorted the crayons one way, ask them to repeat the sorting process using different sorting criteria.

Challenging for older children

Use five circles into which they sort the crayons.

Modifications for younger children

Do this in small groups of three or four. Use a box of sixteen small crayons with the eight jumbo crayons. Use two circles for sorting.

Crayon Eclecticism

Activity

▲ Make a same-and-different chart by writing "Same" on one side of the paper and "Different" on the other.
▲ Show the children the crayons in the boxes.
▲ Ask them to describe how the crayons are the same and how they are different. Record how they are the same on one side of the paper and how they are different on the other.
▲ Slide out the crayons from the boxes and have them add characteristics to the same-and-different lists by examining the crayons.

Same	Different

Interest area
Group time

Science & math principle

Teaches about the diversity of crayons, how there are many different types.

Science & math skill

Encourages communication skills.

Science & math attitude

Develops a desire for knowledge and curiosity.

You need

1 box of 8 jumbo crayons
1 box of 8 glitter crayons
1 box of 8 multicultural crayons
1 box of 16 regular crayons
large sheet of white paper
marker

Extension

Have the children draw designs with each type of crayon and compare the similarities and differences in the way the crayons color.

More challenging for older children

Encourage the children to examine the various crayon boxes and make a same-and-different list of the boxes.

Modifications for younger children

Allow the children to hold, feel and smell the crayons as they make their lists.

Questions you might ask

▲ How many differences did you list? similarities?
▲ If you were a crayon, what color would you be and what size?

Multi-plenty-cation

Science & math principle

Teaches about population through the numbers of crayons.

Science & math skill

Develops observation skills.

Science & math attitude

Encourages curiosity and a desire for knowledge.

▼

You need

2 boxes of 8 regular-size crayons
2 boxes of 16 crayons, one empty and one full
1 box of 48 crayons

Questions you might ask

▲ What are the advantages of having more colors? Are there any disadvantages?
▲ How many boxes of eight and sixteen did it take to make a box of forty-eight?

Extension

Add a box of sixty-four crayons to the mix.

Activity

Read the following poem to the children.

MULTI-PLENTY-CATION
by Sharon MacDonald

A box of brand-new crayons
Were made for kids, you know.
The ends are nice and pointy
And they can scribble fast or slow.

The crayons stand right up
In a neatly colored row.
They slide across the paper.
They just seem to want to go.

There are several colors
From which I like to choose.
I use the reds, the greens,
The yellows and sometimes the blues.

A box of sixteen crayons
Is two times a box of eight.
But what I'd really rather have
Is a box of forty-eight.

Multi-plenty-cation
Is how crayon boxes grow.
Then you can always share them
With all the friends you know.

More challenging for older children

Help the children make an addition chart showing the combination of boxes and crayons it took to make forty-eight.

Modifications for younger children

Share the poem and have the children examine the crayon boxes and the crayons. Let them repackage the crayons and explore the possible combinations.

▲ After you have read the poem, give the children two boxes of eight crayons to examine and count.
▲ Let them see if they will fit inside the empty box of sixteen.
▲ Compare the two boxes to the box of sixteen, looking at the colors and the number of crayons.
▲ Next, show them a box of forty-eight crayons.
▲ Help the children figure out how many boxes, or box combinations, it would take to make a box of forty-eight crayons.
▲ Talk about what they think "multi-plenty-cation" might mean.

Which Crayon Do You Use?

Activity

▲ The children use the paper to take a survey to find out different things.
▲ Encourage them (or help them) survey the other children to find the following information (recording the data on the crayon paper):

    1) their favorite crayon color
    2) the color they use most to draw
    3) the color they wear the most (clothing)
    4) how many boxes of crayons are at home
    5) how many different crayon brands they have at home
    6) what color paper or type of surface is their favorite to draw or to write on: sandpaper, brown paper, drawing paper, cardboard or tissue paper.

▲ Select the "pollsters."
▲ They take a crayon and piece of crayon-shaped paper and interview the other children, gathering the information about one topic (and only one topic) at a time.
▲ They decide how to record the data on the paper.
▲ After the research has been done, they report their results to the class.

Extension

Encourage the children to brainstorm other questions that they can find answers to by polling others, then do it.

More challenging for older children

Encourage the children to find a way to document and record their research on a graphing grid or a chart.

Modifications for younger children

Use only one question.

Interest area
Group time

Science & math principle
Teaches about the diversity of likes and dislikes in the group.

Science & math skill
Encourages communication skills.

Science & math attitude
Develops honesty when interviewing and recording the interviews.

You need

paper, cut in the shape of a crayon
box of 8 crayons

Questions you might ask

▲ Why do you think __ of the children had boxes of crayons at home?
▲ As a result of your poll, what type of paper did most of the children like to draw on?

Interest area
Science and math

Science & math principle

Teaches about a scale when children weigh the crayons.

Science & math skill

Develops children's comparison skills.

Science & math attitude

Encourages children's curiosity.

▼

You need

box of 48 crayons
balance scale
poker chips or lightweight counting chips
graphing grid chart

Questions you might ask

▲ Which one weighs more, the crayon or the poker chip?

▲ How many crayons does it take to balance five chips? ten chips?

▲ What other things could you use to balance the crayons on the scale?

Betting on the Balance

Activity

▲ Using the balance scale, have the children place a crayon on one side of the scale and poker chips on the other, adding as many poker chips as are needed to bring the scale into balance.

▲ They place the crayon (the one in the balance scale) on the first box of the graph and the chip or chips above the crayon.

▲ Proceed with two crayons, three crayons and so on until all the crayons have been used and the chips arranged.

▲ The children can compare how many chips it took for each crayon and discuss why each is different.

Extension

Use lightweight chips made of cardboard or construction paper so the children will have to use more chips to balance one or more crayons.

More challenging for older children

Give each child a chart to record how many chips it took to balance each group of crayons. The child writes the number of crayons or draw circles on the chart to represent the chips.

Modifications for younger children

Do the chart in a group setting with just a few crayons and chips. Place the crayons, chips and scale in the Science and Math Center for the children to explore.

▼

How Many Crayons Make a Line?

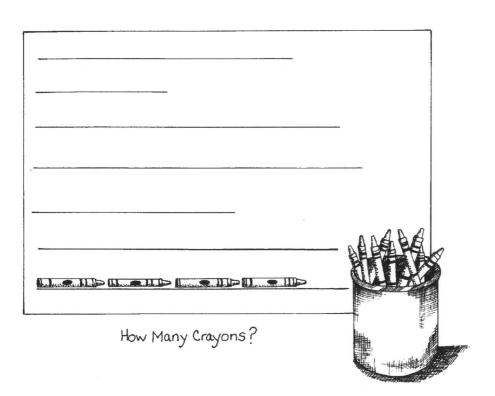

How Many Crayons?

Interest area
Science and math

Science & math principle
Teaches children about patterns that evolve in the lines.

Science & math skill
Develops children's comparison skills.

Science & math attitude
Teaches children to work together and learn cooperation.

▼

Activity

▲ Ask six children to draw lines of different lengths starting at one end of the butcher paper and going as far as they like on the paper.

▲ Space the lines about 6 inches (15 cm) apart.

▲ After the children have drawn the lines, give each of six groups of children a large handful of crayons. They place the crayons end-to-end along the lines to measure them.

▲ Each group measures one line, then counts the number of crayons and reports the number. For example, "Our line is eight crayons long."

▲ Talk about the differences in the lengths of the lines in terms of crayons, then talk about nonstandard and standard measures.

Extension

Measure the length of the lines with a number of nonstandard objects. Compare results to the results with crayons.

More challenging for older children

Use a tape measure to find the length of the line in inches or centimeters as well as crayons.

Modifications for younger children

Use a piece of butcher paper one-foot long and draw three lines. Use jumbo crayons since they are easier to hold.

You need

2 boxes of 64 crayons
sheet of butcher paper, 5' (1.5 meters) long
marker

Questions you might ask

▲ What else could you use to measure the lines?

▲ Why does it take more crayons to measure one line than another?

A Jarring Crayon Prediction

Interest area
Science and math

Science & math principle
Teaches about population when they make predictions and check results.

Science & math skill
Develops comparison skills.

Science & math attitude
Encourages children's curiosity.

▼

You need

tall, thin plastic jar
crayons
adding-machine tape
chart tablet
marker

Questions you might ask

▲ How close was the count to your prediction?
▲ Of the crayons in the jar, which color was there the most of?

Activity

▲ Before the children arrive, draw a number line on adding-machine tape and stand crayons up in the jar so they can see the sides and tops.
▲ Ask the children to estimate how many crayons there are in the jar.
▲ Record all of their predictions.
▲ Ask them to count the crayons by placing them individually on the numbers of the adding-machine tape number line.

Number of Crayons

Sarah – 2

Carolyn – 15

Taylor – 12

Carrie – 8

Peter – 6

Lee – 9

Jorge – 11

Extension

Fill a small box with crayons and put on the lid so the children cannot see how many there are inside. Have them predict how many crayons there are in the box (by shaking it gently). Open the box and ask them to verify their prediction. Close the box. Ask them how many more it would take to fill the box. Encourage them to check their predictions by filling the box with crayons.

More challenging for older children

Place the activity in the Science and Math Center on a tray. Working individually, have the children write their own predictions and check their answers.

Modifications for younger children

Put only three crayons in the jar.

Crayons by Colors

Activity

▲ Before the children arrive, make a three-section graphing grid on a large sheet of white paper.
▲ Open the box of crayons.
▲ Remove the black, white and gray crayons and set them aside.
▲ Glue a red crayon in the first section of the grid, a blue crayon in the second and a yellow in the third.
▲ Put the crayons on a tray for the children to examine.
▲ Ask them to decide in which grid section each remaining crayon in the box should be placed, then ask them why they placed the crayons where they did.

Extension

Add a seventh section to the graph for the neutral colors. Add the gray, black and white crayons to the crayon mix.

More challenging for older children

Use a six-section grid. Add orange, green and purple crayons to the crayon boxes on the grid. Ask the children to place all the remaining crayons in the six sections.

Modifications for younger children

Pick out the crayons that are close to the three primary colors so the children can easily see which remaining crayon goes with which color in each section on the grid.

Interest area
Science and math

Science & math principle
Teaches about patterns as children determine common characteristics.

Science & math skill
Develops communication skills.

Science & math attitude
Develops open-mindedness since there is no "right" way of doing the activity.

You need

box of 48 crayons
large sheet of white paper
marker
glue
tray

Questions you might ask

▲ Why did you place this crayon (pick up one) here?
▲ How many crayons are in each section of the grid?

crayons

Interest area
Science and math

Science & math principle
Teaches about scale as children discover gradation of colors.

Science & math skill
Develops organization skills.

Science & math attitude
Helps children become comfortable with science and math.

You need

box of 8 multicultural crayons
tray
paint color sample strips, in earth tones (available at a paint store)
white paper

Ordering Shades

Activity

▲ Before the children arrive, match the color strips to the crayons (as close a match as possible). Put away the remaining color strips. Each crayon, then, should have a matching color strip.
▲ Have the children color a small area on the white paper and compare it to the color strip, then place the crayon on the matching color strip.
▲ Once matched, encourage the children to put them in order from the darkest to the lightest, then the reverse.

Questions you might ask

▲ Which one is the darkest? lightest?
▲ What other things could you use to compare with the color of the crayon?

Extension

Use red, blue and green crayons from a box of sixty-four and matching color strips.

More challenging for older children

Encourage the children to match the crayon to the color strip without coloring on the sheet of paper.

Modifications for younger children

Since most young children are not developmentally ready to seriate the colors, end the activity when they have matched the crayons to the color strips.

Unwrapping Matchless Crayons

Activity

▲ Before the children arrive, carefully slit the wrappers on the crayons from one entire box of crayons.
▲ Peel off the wrappers.
▲ Smooth out and flatten the crayon wrappers.
▲ Glue them to the cardboard sheet with the printed words showing.
▲ Place the wrapper board and the other box of sixteen crayons on the tray.
▲ Place the tray in the Science and Math Center.
▲ The children match the crayon color to the corresponding wrapper using the word(s) describing the color as a clue.

Questions you might ask

How did you know the crayon matched the wrapper?
What color do you like the best? the least?

Extension

Rather than having wrappers on the board, write the color words on it for the children to match to their crayons.

More challenging for older children

Do the same activity with a box of forty-eight or sixty-four crayons.

Modifications for younger children

Do the same activity with a box of eight crayons.

Interest area
Science and math

Science & math principle
Teaches about properties of crayons.

Science & math skill
Develops observation skills.

Science & math attitude
Encourages children's curiosity.

▼

You need

2 boxes of 16 crayons
cardboard sheet, 5" x 7" (12 cm x 18 cm)
glue
tray

The Crayon Muffin Meltdown

Interest area
Science and math

Science & math principle
Teaches about how crayons change when heated.

Science & math skill
Develops observation skills.

Science & math attitude
Teaches respect for the tools of research.

▼

Activity
▲ The children break the old crayons into small pieces and place the same color crayons in each of the muffin tin cups.
▲ Fill the cups, but not above the edge.
▲ The children will discover what happens when the crayons are melted and then cooled. This can be done as a group or individually.

State the Problem or Question
What will happen when we melt the broken crayons and let them cool?

State the Hypothesis
We believe _____, _____, _____ and _____ will happen when broken crayons are heated in the oven and then allowed to cool.

Method of Research
We will place the muffin tin filled with broken crayons in a preheated oven at 275º F (140º C) for fifteen minutes. We will let the muffin tin cool and then see the results.

Checking the Hypothesis
We will look at our hypothesis and see if we were correct.

Results
We found that the crayons melted and that, when cooled, they were transformed into crayon muffins we could _____ with.

You need
a bucket of old, peeled crayons (discard the brown, black and purple colors)
old muffin tin
oven, preheated to 275º F (140º C)

Questions you might ask
▲ What happened to the crayon pieces?
▲ What can you do with crayon muffins?

Extension
Draw pictures with the crayon muffins.

More challenging for older children
Do the same activity, but mix the colors in the cups of the muffin tin.

Modifications for younger children
Take the crayon pieces out of the oven every five minutes so they can observe the changes. Make sure the children know that the oven and muffin tin are very hot and not to touch either one.

Raising Cups

Activity

▲ Place the cups in the basket.
▲ Put newspaper on the floor in a low-traffic area.
▲ Put the box in the center of the newspaper.
▲ Place the basket of cups and the glue beside the box.
▲ The children glue the cups all over the box, even gluing cups on the cups until every child has contributed to the cup construction.

Extension

Take the construction outside and spray paint it. Have the children vote on a name for the construction and decide on its function. Add stickers labeling the function. For example, if the children named their construction "The Farm," give them animal stickers and have them add to the labeling.

More challenging for older children

Add scissors to the choice of materials and see what happens.

Modifications for younger children

Choose a box with larger surfaces. It is difficult to make the cups stick to each other. If you have a large base, the children have plenty of space to glue cups to the base without having to glue cups on top of cups.

Interest area
Art

Science & math principle
Teaches about interaction as children glue the cups on the box.

Science & math skill
Develops communication skills.

Science & math attitude
Teaches children to be cooperative.

You need

paper and plastic cups, many different sizes and colors
large basket
newspaper
large cardboard box, 2' x 2' x 2' (60 cm x 60 cm x 60 cm)
tacky glue, such as Aleene's (this glue works well on wax-coated and plastic cups)

Questions you might ask

▲ What does the construction look like?
▲ How could you make it different?

Interest area
Art

Science & math principle
Teaches about patterns.

Science & math skill
Encourages communication skills.

Science & math attitude
Teaches children patience.

▼

You need

paper cups (9 or 11 oz.), 1 per child
scissors
ribbon pieces placed in a container
box cutter or an *Xacto* knife
tray

Questions you might ask

▲ How many colors of ribbons did you use?
▲ What would the cup look like if you used only purple ribbon? How would it be different than the cup you created?
▲ What types of ribbon were easiest to use? Hardest?

A Ribbon Weave

Activity

▲ Before the children arrive, use the box cutter to make vertical slits all around the cup, approximately one inch (2.5 cm) apart. Start one-half (13 mm) inch from the top lip of the cup and cut straight down to one-half inch (13 mm) from the bottom. Note that the slits will be closer together at the bottom, due to the tapered shape of the cups.
▲ Place the container of ribbon and the slit cups on the tray.
▲ The children choose a cup, then weave ribbon in and out the vertical slits in the cup.
▲ Ask the children to describe the patterns of ribbons on their cups.

Extension

Use these cup designs for dry snacks at classroom parties, for supplies or for hanging around the room as decoration. With a little modification, they make terrific bells!

More challenging for older children

When collecting ribbon pieces, make sure you select a wide variety of widths. The children might also like to experiment with various sizes of cups.

Modifications for younger children

Do this individually with one-on-one help. This will be a difficult activity for younger children because they lack fully developed motor skills, so make sure the ribbon you use is wide and easy to grasp.

Cups and Soup

Activity

Note: This activity involves boiling water. For safety, have the children move away from their bowls when you add the boiling water and wait several minutes for it to cool before stirring. Soup should be cool to the touch before the children are allowed to eat it.

▲ Open several packages of soup mix and dump contents into the mixing bowl.
▲ Each child dips a teaspoon into the bowl of soup mix and puts the teaspoonful in a small bowl.
▲ Add one-fourth cup of boiling water to each bowl (only the adult does this).
▲ After it cools slightly, the child carefully stirs it with a spoon.
▲ After cooling, give each child a napkin, a spoon and their soup for a snack.

Questions you might ask

▲ Does your soup look like anything else you have had before? What?
▲ What caused the powdered mix to become soup?

Extension

Try different brands of soup mixes to see which mixes the best, which tastes the best and which is the best buy.

More challenging for older children

Taste several instant soups and graph the children's preferences.

Modifications for younger children

Do this activity in small groups that can be well-supervised. Make sure the children understand that the soup is very hot and that it could burn them. Place the hot bowls of soup out of reach of the children until they cool.

Interest area
Cooking and snack

Science & math principle
Teaches about how soup changes as it cooks.

Science & math skill
Develops observation skills.

Science & math attitude
Develops curiosity.

▼

You need

several packages of instant soup in cups (enough for 4 children to share one package)
teaspoon measure
mixing bowl
small bowls, 1 per child
water
pot or kettle for boiling water
stove or hot plate
plastic spoons
napkins

Icebergs in Cups

Interest area
Cooking and snack

Science & math principle
Teaches about the water cycle.

Science & math skill
Develops observation skills.

Science & math attitude
Encourages children to be comfortable with science and math.

▼

You need

16 oz. (500 ml) plastic cup for each child
5 oz. (150 ml) wax-coated paper cup for each child
water
tray
clothespins (optional)
mittens, gloves or hot pads (optional)
apple, grape or orange juice
freezer

Questions you might ask

▲ What caused the juice we added to get cold?
▲ How is the ice cup like a regular cup? How did it happen?

Activity

▲ With help from the children, fill the plastic 16-oz. cups and the smaller 5-oz. paper cups ¾ full of water.
▲ Slide the smaller cup down into the plastic cup. Do this with each child.
Note: If you want the smaller cup to sit squarely inside the larger cup when both are filled with water, use clothespins and attach them on the opposing four sides of the larger cup. The clothespins force the smaller cup to "stand off" from the sides of the larger cup, keeping the small cup in the center of the large one (this allows the water to freeze evenly making a uniform ice cup with the sides of more-or-less equal thickness).
▲ Place all the cups on a tray and put them into the freezer. Leave them about three hours (each freezer is different, so check them every hour to see if the water in the smaller cup is frozen).
▲ Pull out the tray of cups and let them sit for five minutes.
▲ Pull out the small cup from inside the larger one. What you will have left is a partial ice cup frozen inside the larger plastic cup.
▲ Place the small cup cubes in the pouring table and add water for the children to explore.
▲ Let the children examine their partially frozen ice cups (inside the plastic cups). Pour off the about half the excess water that has not frozen. Refreeze the cup for use the following day.
▲ Retrieve the ice cups from the freezer the next day. Pour in juice and let the children have an icy drink.
Note: If you are brave, allow the children to dip their plastic cups in hot water and remove the ice cup from the plastic cup. Check and make sure there are no holes on the sides. Pour in the juice, give them the gloves, mittens or hot pads and allow them to drip as they drink. Very messy, but exciting.

Extension

Do this with several types of juice and graph the children's juice preferences.

More challenging for older children

Encourage the children to draw pictures showing the steps they followed in the ice cup activity and add words (by using invented spelling or dictating the story to you).

Modifications for younger children

Put masking tape on the cups to make them easier to hold (as the ice melts, the outside of the cups get slippery).

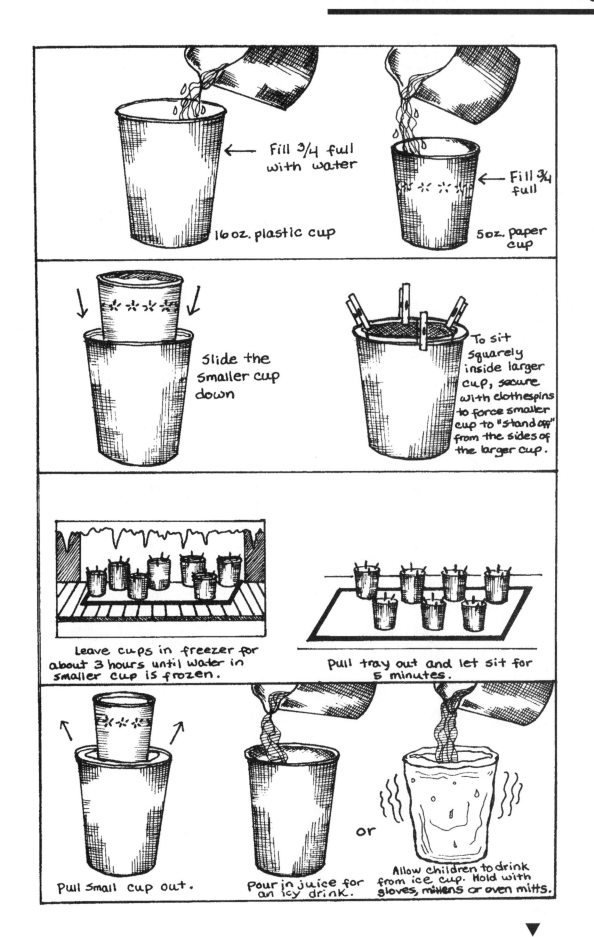

Fill ¾ full with water

16 oz. plastic cup

Fill ¾ full

5 oz. paper cup

Slide the smaller cup down

To sit squarely inside larger cup, secure with clothespins to force smaller cup to "stand off" from the sides of the larger cup.

Leave cups in freezer for about 3 hours until water in smaller cup is frozen.

Pull tray out and let sit for 5 minutes.

Pull small cup out.

Pour in juice for an icy drink.

or

Allow children to drink from ice cup. Hold with gloves, mittens or oven mitts.

Interest area
Cooking and snack

Science & math principle
Teaches about how ginger-bread changes as it cooks.

Science & math skill
Encourages children to compare the mixture to the cooked product.

Science & math attitude
Develops children's patience.

You need

story of the *Gingerbread Boy*
gingerbread mix in a box, each child needs 3 level tablespoons (45 ml) of mix
small pitcher of water
wax-coated paper cups, 1 per child
tablespoon measure
craft sticks, 1 per child
electric skillet

Questions you might ask

▲ Can you describe how you made the ginger-bread?
▲ How do you think you know when the ginger-bread is ready?

Gingerbread Cupcakes

Activity

▲ Read the story of the *Gingerbread Boy*.
▲ After you have read and discussed it with the children, explain that the children are going to make gingerbread in a very unusual way: in cups!
▲ The illustrations show how to cook the gingerbread in individual paper cups. Go through the picture directions and then let the children work independently in work stations.

Station 1
Put 3 (45 ml) level tablespoons of mix in each cup.
Station 2
Add 1 tablespoon (15 ml) of water and stir with a craft stick. (If it seems too dry, add more water.)
Station 3
The adult places four to six cups at a time in the electric skillet that has been preheated to 400° F (200° C). Place the lid on the electric skillet. Cook for 10 to 15 minutes. The cake should spring back at the touch. Set aside to cool.
Note: Since each electric skillet cooks a little differently, you may want to test the baking time of your skillet at home before attempting this activity at school.
Station 4
Sit down and eat the gingerbread.

Extension

Make gingerbread cookies and compare the taste of cookies with the taste of cup-baked gingerbread.

More challenging for older children

Let the children organize the stations and make their own picture directions for this activity.

Modifications for younger children

Do this activity in small groups, and omit the work station format. You will have to help the group perform certain tasks. Mark off a square area with red masking tape around the electric skillet, making the area and the skillet inside a "no touch" zone.

Station 1

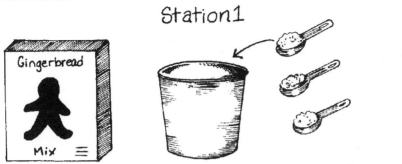

Put 3 level tablespoons of mix in each cup.

Station 2

Add one tablespoon of water and stir with craft stick.

Station 3

Adult places 4 to 6 cups at a time in the electric skillet that has been preheated to 400 degrees.

Station 4

Eat!

The Ping-Pong Ball & Cup Toss

Interest area
Gross motor

Science & math principle
Teaches about cause and effect, depending on how the ball is thrown.

Science & math skill
Develops observation skills.

Science & math attitude
Develops curiosity.

▼

You need

5 plastic or paper cups, 9 or 11 oz. (250 ml to 325 ml)
permanent marker
tacky glue, such as Aleene's
15" x 6" (37 cm x 15 cm) sheet of heavy cardboard
21 Ping-Pong balls in a small basket
masking tape

Questions you might ask

▲ What other way could you get the balls in the cups?
▲ Pretend you are the Ping-Pong ball. How would you feel? Into which cup would you like to be tossed?

Activity

▲ Before the children arrive, write the numbers 1 through 5 on the cups with the permanent marker.
▲ Glue the cups in a line (in numerical order) with their centers 3 inches (8 cm) apart on the heavy cardboard sheet (see illustration) or an empty fabric bolt board.
▲ Place the board and the container of Ping-Pong balls in a low-traffic area for the children's use.
▲ Establish a toss line 2 or 3 feet (60 cm to 1 m) from the board. Mark it with masking tape.
▲ The children toss the balls at the cups with the goal of getting one ball in the #1 cup, two balls in the #2 cup, three balls in the #3 cup, and so on.

Extension

Number the balls 1 through 5. The children must toss the balls in the matching numbered cups from one to five.

More challenging for older children

Encourage the children to come up with a way to keep score. They could also change the game, adding more cups and Ping-Pong balls.

Modifications for younger children

Use a nerf ball and large boxes.

What Fits in a Cup?

Activity

▲ Line up the cups on the tray.
▲ Place the items, including bottles, in the large basket.
▲ Have the children predict which items will fit into which cups, then have them place the items in front of that cup.
▲ After the predictions have been made, have them test their predictions by putting the items into the cups.

Note: Many items will fit into several of the cups. What is important is that the children make choices, then predict, test and learn from the choices. There is no "right" answer.

Questions you might ask

▲ Can you place more than one item in a cup?
▲ What would happen if the cup were on its side? Could you fit bigger items into it?

Extension

Put out one cup and lots of items. Have the children decide first if an item will fit into the cup, then test it. If it fits into the cup, they put it on a tray with a sign that says "YES." If it does not fit into the cup, they put it on a tray that says "NO."

More challenging for older children

Put out more items than you have cups. This will give the children more problem solving to do. They may look for items that fill a container more completely or for the container that holds the most items, or the least.

Modifications for younger children

Put out four very different-sized cups and four different items that will fit into each cup.

The Cup-Tapping Song

Activity

▲ The children tap the cups together to the song below.

▲ Select a different child each time the song is sung to decide how the group will tap their cups together.

▲ The tune is "My Name is Stegosaurus." If you do not know the tune, just make up your own. The children will follow right along.

CUP TAPPING SONG

by Sharon MacDonald

These cups are made for tapping.
It's a funny little game I play.
I tap them here, I tap them there.
Come play my funny game this way.

Tap in and out and in and out.
Tap down and down and down.
Tap over, under, on your head.
Tap all your body 'round.

Questions you might ask

▲ What tapping motion makes the loudest noise? the softest?

▲ How are the sounds of the tapping the same? different?

Extension

Tap the cups to other songs.

More challenging for older children

One child creates a tapping pattern and another child adds to the pattern.

Modifications for younger children

Use smaller cups and tape the cups to make them easier to hold. Use simple rhythmic patterns.

Cup Classes

Activity

▲ Ask the children to place the different cups on the window-shade grid, graphing the cups by type, color, size or shape.

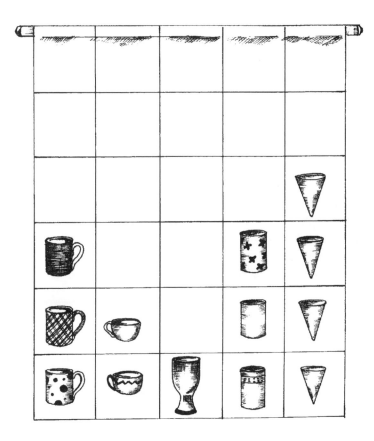

Interest area

Science and math

Science & math principle

Teaches children to discover the properties of cups.

Science & math skill

Encourages communication skills.

Science & math attitude

Develops curiosity.

▼

You need

a wide variety of cups (e.g., cone-shaped cups, cups with and without handles, medicine cups, glass cups, wooden cups, measuring cups, cups with saucers, cups with lids and straws, and paper cups). The cups should be different colors, sizes and shapes and made from paper, plastic, glass or wood.

window-shade graph from Rock Graphing, page 197

Questions you might ask

▲ How are all the cups alike? What would you put in them to drink?

▲ Which size (shape, color, etc.) do you like the best? Why?

Extension

Graph the cups by the material from which the cups are made or where the cups can be found (dentist's office, restaurant, carry-out restaurant, home).

More challenging for older children

Graph the cups by their function (for example, cups that are used for measuring or drinking).

Modifications for younger children

Serve the children juice using as many different kinds of cups as possible. When they finish, help them place the cups on the graphing grid by cup color. At another time, help them graph the cups by size, then later by shape.

How Many Cups Are Cupped?

Interest area
Science and math

Science & math principle
Encourages children to examine the cups, learning about population.

Science & math skill
Encourages children to compare the estimates and actual number of cups.

Science & math attitude
Develops children's curiosity.

▼

You need

10 to 12 paper or plastic cups graduated in size from 5 to 32 oz. (150 ml to 1 L) capacity
tray
index cards
chart tablet and marker

Questions you might ask

▲ What is another way we could find out how many cups are inside the large cup?
▲ How are the cups different? the same?

Activity

▲ Before the children arrive, write each child's name on a separate sheet of paper.
▲ Stack all the cups inside the large one.
▲ Place the stacked cups on a tray.
▲ The children are going estimate how many cups are inside the large cup.

State the Question
How many cups are in the large one?

State the Hypothesis
Each children will be able to establish a method to estimate and determine how many cups are inside the large one. Each child will make a prediction. Each child's prediction will be written on a chart beside the child's name.

Method of Research
A small group of children will take the cups from the larger one and count them. They will write the results on an index card and then report to the rest of the class how many cups were in the larger one. The children may use a number line to count their cups so they learn what the number looks like and how to write it. They will write the number on the index card.

Checking the Hypothesis
We will look at our list of predictions and see if we were correct.

Results
We found ____ cups inside the large cup.
_____children guessed the right number.
_____children were one or two numbers from getting the correct result.
_____children were more than two numbers from getting the correct result .

Extension

Fill a four- or five-ounce cup with jelly beans. The children guess how many jelly beans it took to fill the cup. They check their predictions by counting the jelly beans. Give each child a cup full of jelly beans and have them count as they eat. Compare the actual count with their predictions.

More challenging for older children

Place all the cups inside the largest cup and turn the largest cup upside down so the children cannot see the top rims of the cups. Then have the children estimate how many cups are inside.

Modifications for younger children

Start with three or four cups. As they become better estimators, add more cups.

Around the World Rice

Activity

Note: *Everybody Cooks Rice* by Norah Dooley is an excellent story. The book shows how families from many cultures cook rice. The story is an excellent introduction to this activity. *Everybody Cooks Rice* also introduces the children to the different utensils, or tools, used to accomplish the same task of eating. It suggests to the children that different tools can be used to accomplish the same thing.

▲ Cook the rice according to the package directions. Use short-grain white rice so it will be sticky when cooked.

▲ Put a scoop of cooked rice in a bowl for each child.

▲ Encourage the children to try eating with chopsticks. This may be difficult for some children, so let them switch to plastic spoons if they would like.

Questions you might ask

▲ What have you eaten before that contains rice?

▲ How did the rice change during cooking?

Extension

Cook other types of rice and compare.

More challenging for older children

Examine and compare uncooked white rice, brown rice, medium-grain rice, long-grain rice and Asian dark rice.

Modifications for younger children

Put chopsticks in the Home Living Center for a week or two before doing this activity with younger children. Let them practice with them. If they cannot manage the chopsticks, demonstrate how to use them and talk about how different tools can be used to do the same thing.

Interest area
Cooking and snack

Science & math principle
Teaches children about the properties of rice, cooked and uncooked.

Science & math skill
Encourages observation skills.

Science & math attitude
Develops children's curiosity.

▼

You need

Everybody Cooks Rice
package of short grain white rice
water
pot
stove or hot plate
chopsticks for each child
small bowl for each child
plastic spoons
ice cream scoop

Bread and the Little Red Hen

Interest area
Cooking and snack

Science & math principle
Teaches children about change as the flour is made into bread.

Science & math skill
Develops observation skills.

Science & math attitude
Develops children's curiosity.

You need

story of *The Little Red Hen*
1 package dry yeast
honey
vegetable oil
salt
warm water
whole wheat flour (add the flour you ground in The Wheat Grind, page 122)
measuring spoons
measuring cup
2 bowls
stirring spoon
towel
2 bread loaf pans, greased and floured
oven

Activity

▲ Read the story of *The Little Red Hen*. After you have read and discussed the story, explain that they are going to be "little red hens" and make bread.
▲ Use the picture (rebus) directions to show the children how the ingredients are used and the process by which they are added to make bread.
▲ In the first bowl, dissolve the yeast in 2½ cups (625 ml) of warm water; add 3 tablespoons (45 ml) of honey and 3 tablespoons (45 ml) of oil to the yeast and warm water mixture. Set the bowl aside.
▲ In the second bowl, mix 6 cups (1.5 liters) of flour and 2 teaspoons (10 ml) of salt. Pour the contents of the first bowl into the second. Stir. When all the liquid is absorbed by the flour, knead the mixture.
▲ Cover the dough with a towel. Set aside in a warm place. The dough will double in size in 2½ to 3 hours.
▲ Divide into two equal balls. Knead the dough and place into the two bread loaf pans that have been greased and sprinkled with flour.
▲ Bake at 350° F (180° C) for 40 to 45 minutes. Remove from the oven and set aside to cool for 30 minutes. Turn out the bread loaves to finish cooling.
▲ Serve slices to the children to eat for snack.
Note: I like this recipe because the bread seems always to "survive," no matter what the children do when mixing or kneading it.

Questions you might ask

▲ Describe the sequence of events that happened to make the flour change to bread. What happened first? second? third?
▲ What ingredients can you name?

Extension

Make other types of breads and compare them.

More challenging for older children

Ask the children to draw pictures or use written words to retell the sequence for making bread.

Modifications for younger children

Involve the children as much as possible in the making and baking process. Work with small groups of three to four children at a time so they can each have a better view.

Little Red Hen's Bread

you need:

one package dry yeast

2½ cups of warm water

3 tablespoons of Honey

3 tablespoons of oil

2 teaspoons of salt

6 cups of whole wheat flour

1 teaspoon

one tablespoon

one measuring cup

Dissolve 1 package of yeast in 2½ cups of warm water.

Add 3 tablespoons of honey and oil to the liquid mixture.

In another bowl mix 6 cups of flour and 2 teaspoons of salt.

Pour liquid mixture into the flour mixture. mix well.

Knead.

Cover the bowl with a towel. Set aside until the dough has doubled in size.

Punch down the dough and divide into 2 balls. Knead. Place in 2 loaf pans.

Bake at 350° for 40 to 45 minutes. Remove and cool.

Slice the bread.

Eat.

Interest area
Cooking and snack

Science & math principle
Teaches children about change as they make tortillas.

Science & math skill
Develops observation skills.

Science & math attitude
Develops children's curiosity.

You need

large bowl
mixing spoon
measuring cup
measuring spoons
all-purpose flour
salt
baking powder
vegetable shortening (such as Crisco)
warm tap water
rolling pin
electric skillet, comal or portable burner
spatula
fork
paper plates

Making Flour Tortillas

Activity

▲ Mix 3 cups (750 ml) of flour, 1 teaspoon (5 ml) of salt and ⅓ teaspoon (1.5 ml) of baking powder in a large bowl.

▲ Using a fork, blend in 1 heaping tablespoon of shortening. Mix until the flour mixture looks like small pebbles.

▲ With your hands, gradually mix in 1 cup (250 ml) of warm water (the texture will be like dough but not sticky).

▲ Divide the tortilla dough into balls the size of golf balls. Put flour on the work surface and all over the children's (washed) hands.

▲ They roll out the balls, then pat them flat. Give them a floured rolling pin and have them roll the tortillas as flat as they can or use a tortilla press.

▲ Place the tortilla on the hot comal or electric skillet (preheat to 400° F or 200° C).

Note: A comal is a flat cast-iron skillet with no sides. You can also use an electric skillet or a cast-iron skillet.

▲ Cook until brown on one side, then use the spatula to turn the tortilla to cook on the other. If the tortilla puffs up, use the edge of the spatula to make a small perforation in the center of the raised surface.

▲ Set the tortillas aside to cool for about three minutes. Eat them for snack.

▲ A special thanks to Eva Gutierrez, San Antonio, Texas for this exceptional, traditional recipe.

Questions you might ask

▲ How is the tortilla like a piece of white bread?

▲ What could you roll up in your tortilla to make a sandwich?

Extension

Compare the taste of the tortillas made in class with the ones purchased at the store.

More challenging for older children

Ask the children to retell, with pictures or words, the sequence involved in making the tortillas.

Modifications for younger children

Put a red masking tape box around the hot skillet or burner to define the "no touch" zone. Help them to make the dough balls and to use the rolling pin. Note: I have found it helpful to get the dough ball partially rolled out with the pin. Then I place my hands over the child's partially completed tortilla to demonstration the rolling motion. I let them finish on their own.

Warm tap water

Comal

Spatula

Mix 3 cups flour, 1 teaspoon salt,
and 1/3 teaspoon baking powder

Blend in one
heaping tablespoon of shortening

Gradually mix 1 cup of
warm water with hand

make balls

Interest area
Cooking and snack

Science & math principle
Teaches children about irreversible change as the rice pudding cooks.

Science & math skill
Develops observation skills.

Science & math attitude
Encourages children's curiosity.

▼

You need

2 cups (500 ml) cooked rice
1¹/₃ cups (325 ml) milk
¹/₈ teaspoon (.5 ml) salt
5 tablespoons (75 ml) sugar
1 tablespoon (15 ml) soft margarine
1 teaspoon (5 ml) vanilla
3 eggs
1 teaspoon (5 ml) lemon juice
vegetable shortening
sugar cookies
resealable bag
large spoon
9" x 13" (23 cm x 32 cm) baking pan
aluminum foil
measuring cups
measuring spoons
rolling pin
oven

Rice Puddin'

Activity

▲ Preheat the oven to 325° F (160° C).
▲ The children help you put the cooked rice, milk, salt, sugar, margarine, vanilla and eggs into a bowl.
▲ Beat all of the ingredients well with the spoon.
▲ Add lemon juice to the mixture and stir.
▲ Grease the baking pan with vegetable shortening.
▲ Place 12 sugar cookies in a large resealable plastic bag. The children use the rolling pin to crush them into small crumbs.
▲ Cover the bottom of the baking dish with cookie crumbs. Pour in the rice mixture. Add more cookie crumbs on top.
▲ Bake for 50 minutes. Serve when it is golden brown on the top, or let cool and serve cold. Top each serving with marmalade or jam if desired.

mix 2 cups rice, 1 1/3 cup of milk, 1/8 tsp. salt, 5 tablespoons of sugar, one tablespoon of margarine, and one tsp. vanilla

Add three eggs

Add and stir in 1 tsp. lemon juice

grease the baking dish

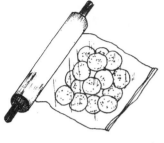

place 12 cookies in a baggie and crush

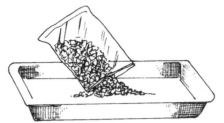

cover bottom of pan with ½ of the cookie crumbs

pour in rice mixture

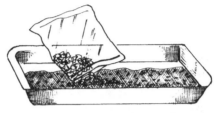

cover the top with the rest of the cookie crumbs

Bake for 50 minutes at 325°

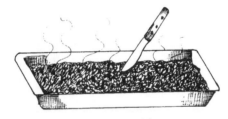

slice and eat!

Questions you might ask

▲ How did the rice change?
▲ What do you like about rice pudding? dislike?

Extension

Make a list of why they think people use rice as a dessert.

More challenging for older children

Ask the children to illustrate or write a recipe book showing each step in the process of making rice pudding.

Modifications for younger children

Involve the children in the preparation process with as little help as necessary. If they have trouble stirring, give them a large pronged fork so there is less resistance when they stir.

The Wheat Grind

Interest area
Group time

Science & math principle
Teaches children about irreversible change as they grind the wheat.

Science & math skill
Teaches children that observing carefully is a useful tool to learn things.

Science & math attitude
Encourages children's desire for knowledge (curiosity).

▼

You need

whole wheat grains
measuring cup
mortar and pestle (in Spanish it is called a *Molcahette*)
tray

Questions you might ask

▲ How did the wheat change? Why?
▲ What are some things that are made from the flour we get from wheat?

Activity

▲ Place ¼ cup (60 ml) whole wheat in the mortar.
▲ The children take turns crushing and grinding the wheat into flour.
Note: save the flour for bread making with the children.

Extension

Crush and grind other grains: oats, wheat, rye and rice. Compare the flours and cook snacks with them.

More challenging for older children

Try using other tools to grind the wheat, like a rock, hammer and potato masher. Talk about why they think the mortar and pestle were invented by people to do the job of crushing and grinding wheat.

Modifications for younger children

Place a tablespoon of wheat in the mortar. It will take time for younger children to crush and grind before change in the wheat can be seen. Many of the children will not be able to make the pestle work in the mortar because they lack the strength to operate the tool. They can, however, observe others and grasp the change taking place in the wheat. Try letting them grind a wheat cereal first as practice.

Puffs or Stalks: Wheat Is What?

Activity

▲ Pass around a handful of puffed wheat.
▲ Talk about what puffed wheat looks like before it is puffed (it is popped much like popcorn).
▲ Encourage children to taste it.
▲ After tasting it, examine the stalk of wheat.
▲ Show the children the other grains in the baby food jars.
▲ Make a list with the children of how all of the grains are the same and how they are different.

Questions you might ask

▲ Which grain is cooked and which is raw?
▲ How are they alike and how are they different?

Extension

Use an electric skillet and cook the wheat, puffing your own wheat.

More challenging for older children

Seriate the jars of wheat grain by the degree of hardness of the grains.

Modifications for younger children

Do this activity in groups of three or four. Put a teaspoonful (5 ml) of each grain on a small plate. Have them feel, examine and taste the grains.

Interest area
Group time

Science & math principle
Teaches children about the properties of wheat and wheat products.

Science & math skill
Teaches children to observe similarities and differences.

Science & math attitude
Develops a desire for knowledge.

▼

You need

small bag of puffed wheat (in the cereal section of the grocery store)
stalk of wheat (this can be purchased at florist or craft shop)
baby food jar filled with cracked wheat
baby food jar filled with whole wheat
baby food jar filled with whole wheat flour
chart tablet and marker

Wheatfest

Interest area
Group time

Science & math principle
Teaches about the diversity of wheat products.

Science & math skill
Develops communication skills.

Science & math attitude
Encourages children to keep an open mind and be able to change their mind.

▼

You need

whole wheat bread
whole wheat crackers
flour tortillas
shredded wheat cereal
pretzels
paper plates
different colors of stickers
chart tablet and marker

Questions you might ask

▲ How are all the products the same? different?
▲ What is the best thing about wheat? Why?

Activity

▲ Make a chart on the chart tablet paper similar to the illustration.
▲ Before the children arrive, place small amounts of each wheat product on a plate to be tasted by the children.
▲ Post the chart in the classroom.
▲ Talk about each wheat product and taste it.
▲ Ask the children to decide which two products they like the best.
▲ Give each child two stickers, each one a different color. Have them come to the chart and put their stickers in the column of their favorite choices.
▲ When all the children have done this, add up all the stickers to determine the top two favorites.
▲ Serve those two grain products for snack on the following two days.

Extension

Ask the children to bring other wheat items to taste and record on the chart.

More challenging for older children

Give the children a sheet of paper with each product listed on it and have them rank their taste preferences from one to five with number one being their favorite.

Modifications for younger children

Use only two or three products for tasting. Later, taste others. Have them choose only one favorite rather than two.

▼

Rice at Home

Activity

▲ Write a letter to the parents explaining your study of grains. Let them know that you will be sending home index cards and hope that they can write down a home recipe for cooking rice.

▲ When the children return with the recipes, select four from the stack and prepare a chart.

▲ Make two columns. Name the recipe on one side of the chart and list all the ingredients and processes that are the same about each on the other side.

▲ At another time during the day, or on another day, repeat the activity until every recipe has been compared. If similar recipes are sent in, do not duplicate ingredients and processes.

▲ Review the final list with the children. The children will be surprised how short the ingredient list is and how similar the processes are.

▲ After completing the activity, put all the recipe cards in a recipe box. This box can be placed in the Home Living center for the children to use when "cooking."

Science & math principle

Teaches about the diversity of rice recipes.

Science & math skill

Encourages children to compare recipes.

Science & math attitude

Encourages children to be curious about different ways to cook rice.

▼

Extension

Vote on the recipe the children would like to cook, then cook it.

More challenging for older children

Ask the children to work in groups to write the recipes in a book and illustrate it. Send home the recipe book as a gift to the parent(s).

Modifications for younger children

When you write the parent letter, ask for a simple recipe. Rather than making a list of all the ingredients and processes, just talk about them.

You need

several index cards for each child
large chart and markers

Questions you might ask

▲ What ingredient is in every recipe besides rice?
▲ How many of these rice meals have you eaten?

The Lure of Labels

Science & math
principle

Teaches about the properties of different boxes.

Science & math
skill

Develops observation skills.

Science & math
attitude

Encourages children to be curious.

▼

You need

different boxes and packages with wheat ingredients like cake mix, pasta, crackers, cereal and those without wheat as an ingredient
large basket
large tray divided down the middle with colored masking tape
small sign that says "wheat" and one that says "no wheat"

Activity

▲ Place all the food products in the large basket beside the divided tray.
▲ Place the "wheat" sign on one side of the tray and "no wheat" on the other side.
▲ The children choose a box or a package, "read" the ingredient label and place it on the tray under the appropriate sign.
▲ Talk to the children about the ingredients in the box. Explain that the ingredients are listed on the outside of the box in order of their quantity inside. In other words, the list of ingredients is a seriation itself, with the ingredients written in order from the largest amount to the smallest.

Note: If they have trouble reading the word "wheat," post it in big letters near the labeling activity.

Questions you might ask

▲ Is the word "wheat" at the beginning, in the middle or at the end of the label? What does that mean?
▲ Why do you think the people who made the product put the ingredients on the outside of the package for you to see?

Extension

Ask the children to check all cans, boxes and frozen food products to find the word "wheat." Have the children bring in a list of all the things they found with the word wheat on the box or package. Compare the lists that the children bring.

More challenging for older children

Ask them to bring in boxes and packages with wheat as an ingredient to examine at school with other children. Graph the number of boxes and packages brought in by product type (cake mixes, cereal, crackers, pasta and "other").

Modifications for younger children

Use three or four boxes. This activity will help younger children learn that the words on the box tell us what is in the box. Some will be able to find the word wheat, but others will not.

Note: The goal of this activity is for younger children to learn that words represent things and that they can look for things by "reading" (they are learning that they are able to solve some of their problems on their own).

Pennies and Honey Buns

Activity

Sing and dramatize the following song with the children.

HONEY BUNS

(tune: Six Little Ducks)

Five little Honey Buns in a baker's shop
Big and round with sugar on the top.
Along came _____ with a penny one day.
He (she) bought a Honey Bun and took it on his (her) way.

Four little Honey Buns in a baker's shop
Big and round with sugar on the top.
Along came _____ with a penny one day.
She (he) bought a Honey Bun and took it on her (his) way.

Three little Honey Buns in a baker's shop
Big and round with sugar on the top.
Along came _____ with a penny one day.
He (she) bought a Honey Bun and took it on his (her) way.

Two little Honey Buns in a baker's shop
Big and round with sugar on the top.
Along came _____ with a penny one day.
She (he) bought a Honey Bun and took it on her (his) way.

One little Honey Bun in a baker's shop
Big and round with sugar on the top.
Along came_____ with a penny one day.
He (she) bought a Honey Bun and took it on his (her) way.

No little Honey Buns in a baker's shop
Big and round with sugar on the top.
Along came _____ with a penny one day.
No Honey Buns for her (him) today.

Interest area

Music and movement

Science & math principle

Teaches about number patterns.

Science & math skill

Develops observation skills.

Science & math attitude

Encourages children to work together (cooperation).

▼

You need

individually wrapped honey buns (1 for every 2 children)
paper plates
plastic knives (1 for every 2 children)
6 pennies
tray

▲ Give each of six children a penny.

▲ Place five honey buns on a tray, each on a separate paper plate.

▲ Choose a child to be the "baker," selling the honey buns by displaying them on the tray.

▲ As the children in the class sing the song, the children with pennies act out the stanzas by coming up and purchasing a honey bun, one at a time. These buns will be eaten later. When the last child with a penny tries to buy a honey bun, she finds that none are left.

Note: The first time you sing and dramatize the song, I recommend that you be the one with the sixth penny who tries to buy a honey bun, but cannot because none are left. You can model an appropriate response so when it is a child's turn, he will be able to manage his disappointment more constructively.

▲ Repeat this activity and the song until every two children have a honey bun.

▲ Each pair of children unwraps and cuts their honey bun in half, so each child gets half a honey bun.

▲ Talk with the children about part/whole relationship, or fractions, depending on their level.

Questions you might ask

▲ What are all of the things you think a baker might put in a honey bun?

▲ How many people shared a honey bun?

▲ What are other ways to divide (and share) a honey bun?

Extension

Cook honey buns with the children.

More challenging for older children

Ask older children to cut their honey buns in halves, thirds and fourths.

Modifications for younger children

Most of the children will have trouble waiting to eat their honey bun, so cut one of the buns into bite sizes. The children can have a taste first, knowing they will have more when the activity is completed.

The Big Sift

Activity

▲ Place the whole wheat flour, cracked wheat and whole wheat in the dishpan with different sized measuring cups, a sifter and a scoop.

▲ The children estimate how many scoops are needed to pour in the sifter (crank or squeeze type) to fill the cup they have selected beneath the sifter.

▲ The children then scoop and sift the flour into the cup. (Sifted flour takes up more space than unsifted flour because of the added air.)

Note: Save this flour mixture to make playdough.

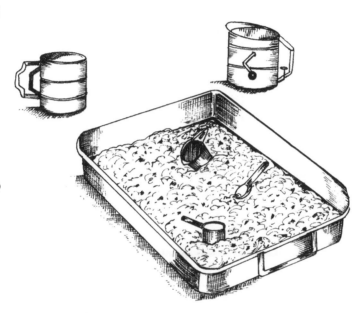

Extension

Compare the whole wheat flour sifting to regular all-purpose flour sifting.

More challenging for older children

Put out a variety of scoops for the children. Have them guess which scoop will hold exactly enough flour to fill a specific measuring cup beneath the sifter. For example, a large scoop may hold the right amount of presifted flour to fill ¾ cup (175 ml) when sifted.

Modifications for younger children

The "crank" sifter is easier to use. For those who have difficulty making their two hands work together (as is needed when using a crank sifter), use a tea strainer. Have the children pour in the flour and shake it.

Interest area
Sand and water

Science & math principle

Teaches children to observe the interaction between the sifter and the flour.

Science & math skill

Encourages children to compare the different amounts.

Science & math attitude

Teaches children to be comfortable with science and math by using a new tool.

▼

You need

leftover flour, cracked wheat and whole wheat from Puffs or Stalks: Wheat Is What?, page 123
dishpan
"squeeze" flour sifter
"crank" flour sifter
scoop
measuring cups

Questions you might ask

▲ How does the flour change when you sift it?
▲ What happens to the whole wheat when you sift it?

Which Bread Turns Your Head?

Interest area
Science and math

Science & math principle
Teaches about patterns as children try to find similarities.

Science & math skill
Teaches children to organize information by graphing it.

Science & math attitude
Encourages children's curiosity.

▼

You need

Bread, Bread, Bread by Ann Morris

the window shade graph from Rock Graphing, page 197

variety of breads: pita, bread sticks, French bread, a bun, a roll, a tortilla, a cracker, a bagel, a pretzel, white bread and fry bread

clear plastic wrap

large tray

Questions you might ask

▲ How are the bagel and the bun alike? different?

▲ Do you have more long bread than short bread? more round than square?

▼

Activity

▲ Before the children arrive, wrap in clear plastic any breads that are not already packaged in a clear wrapper. Tape closed.

▲ Read and talk about Ann Morris' book, *Bread, Bread, Bread*.

▲ Place all the bread types on the tray.

▲ Encourage the children to decide how to graph the bread on the window-shade grid. They place the breads on the grid. They could do it by size, shape, color, hardness, length or texture.

Note: Use the bread for a snack on the day of the activity or the day after.

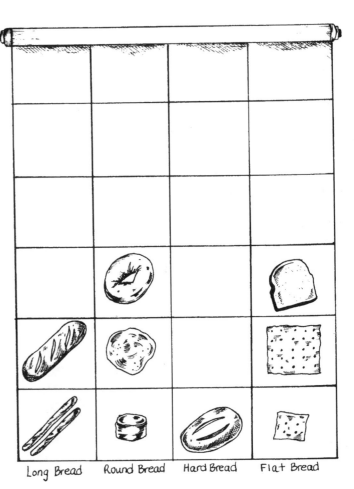

Long Bread   Round Bread   Hard Bread   Flat Bread

Extension

Ask the children to look for a different kind of bread each time they go to the grocery store or have the children bring in different breads for graphing and tasting.

More challenging for older children

Draw a Venn Diagram and have the children sort the bread by one charac-teristic for one of the circles and by another characteristic for the other circle. The breads with both of the characteristics go in the middle of the diagram where the circles overlap.

Modifications for younger children

Use three or four bread items. Make this an introductory activity for the children, where they just look at and describe the different breads.

Just Plain Grain

Activity

▲ Make a sorting tray by sectioning off the tray with colored tape. For this activity, make seven sections on the tray.

▲ Pour all of the grains in the small container.

▲ One section is for the container holding all the grains, the other sections are for the children to sort the grains by type using the tweezers.

▲ The children use tweezers to pick up a grain and place it in a section on the tray.

▲ They place all similar grains in the same section. Continue until all of the grains have been sorted.

Extension

Graph the grains by type.

More challenging for older children

Make a list of all the characteristics that are alike about each grain, then list all the things that are different.

Modifications for younger children

Using the tweezers is hard for most younger children. Have them use their fingers. Rather than placing all of the grains in a small container, use a second tray so the grains are spread out and it is easier to see the differences.

Interest area
Science and math

Science & math principle
Teaches children about the different kinds of grains (diversity).

Science & math skill
Teaches children to organize what they have learned.

Science & math attitude
Encourages children to be comfortable with science and math as they use tools.

▼

You need

10 oat grains
8 wheat grains
8 rye grains
6 barley grains
4 wild rice grains
3 white rice grains
small container
tweezers
tray
colored tape

Questions you might ask

▲ Which grains have the most? least?

▲ How are the grains alike? different?

A Toasted Story

Science & math principle

Teaches about cause and effect when children toast their bread.

Science & math skill

Encourages observation skills.

Science & math attitude

Teaches children a respect for science.

▼

You need

large jar (a pickle jar is good)
candle, 2" (5 cm) in diameter and 3" (8 cm) tall
bamboo skewers
scissors
long fireplace matches
thick-sliced bread
cutting board and knife
pop-up toaster
paper plates

Questions you might ask

▲ How did the bread change? (Point out that once bread is cooked, it is called toast.)

▲ By which method do you think most people toast bread today ? Why?

Activity

▲ Before the children arrive, snip off the sharp ends of the bamboo skewers.

▲ Place the candle inside the jar, in the center.

Note: Talk to the children about handling the jar carefully, since it is easy to knock over the candle. If you are uncomfortable with that possibility, use *Sticky-Tac* and adhere the candle to the bottom of the jar.

▲ With close supervision, let the children cut some of the thick white bread into 1- to 2-inch (2.5 to 5 cm) cubes, then push the cubes onto skewers.

▲ Light the candle. The children toast their bread over the candle (since the candle flame is six to nine inches away from the bread, the bread will not be burned). When their bread cubes are toasted on all sides, place them on the paper plates.

▲ Move to the pop-up toaster. Insert two slices and toast the bread. Cut the toasted bread into 1- to 2-inch (2.5 to 5 cm) cubes and add it to their plates. When each child has both kinds of toast, have a "toast tasting."

Note: I have cooked many foods during the school year using a candle in a pickle jar. I use a hot-glue gun and glue the pickle jar to an 18-inch (45 cm) square piece of plywood. This prevents the children from knocking over the jar.

State your Problem or Question

Which method will cook toast the best (the most evenly and the most quickly), an open candle flame or a toaster?

State the Hypothesis

We think the toast will taste the best using the _____ and cook the quickest using the _____.
(This can be an individual or a class prediction.)

Method of Research

We will cook bread over an open candle flame and with a pop-up toaster. We will taste each. We will decide which tastes the best, and which is the easiest.

Checking the Hypothesis

We will look at our predictions and see if we were correct.

Results

We found that toast cooked _____ tasted the best; and that toast cooked the quickest_____.

Extension

Cook other foods over the open flame like marshmallows, buttered pecans, mushrooms and sections of squash.

More challenging for older children

Add a third way of toasting bread by using the oven. Compare all three methods: look at taste, texture, ease of cooking and the evenness of the toasting process.

Modifications for younger children

Supervise this activity well. Have an oven mitten ready for them to use when removing their hot toast from the skewer.

Lemonade Cubes

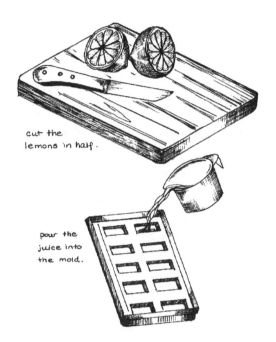

cut the lemons in half.

Twist the lemon to get the juice.

pour the juice into the mold.

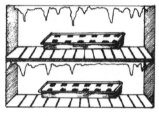

Freeze.

Interest area
Cooking and snack

Science & math principle
Teaches about change as children make the lemon cubes and the lemonade.

Science & math skill
Develops observation skills.

Science & math attitude
Teaches respect for the tools used.

You need

10 lemons
cutting board and knife
small ice cube trays or
  several candy molds
lemon juicer
pitcher
paper cups
sugar cubes
small child-size pitcher of
  water
popsicle sticks

Questions you might ask

▲ What do you think will happen to the lemon juice in the mold when we put it in the freezer?
▲ How is freezing the lemon juice going to make it different?

Activity

▲ Cut the lemons in half.
▲ The children squeeze the lemon halves into a pitcher using the lemon juicer.
▲ Pour the lemon juice into the candy molds, enough so that each child will have two lemon ice cubes. Freeze the molds.
▲ When you are ready to do the activity, set up lemonade-making-and-drinking stations as follows.

Station 1
The children put two cubes in their cups.
Station 2
The children fill their cups with water using the small pitcher.
Station 3
The children place one or two sugar cubes in their cups; they stir their lemonade with the popsicle sticks.
Station 4
The children sit down and drink their lemonade (they know it is ready to drink when the lemon ice cubes have melted).

Extension

Do this with other fruits like limes, grapefruits and oranges.

More challenging for older children

Encourage the children to figure out what needs to be done at each lemonade station and to create the stations. They might invite another class to come and share their lemonade.

Modifications for younger children

Depending upon the age of the children, squeezing the lemons may be very difficult and time consuming. Squeeze all the lemons ahead of time, except for one. Demonstrate the process or allow the children to help squeeze the lemons.

Frozen and Hot Foods

Activity

▲ Make a chart like the one in the illustration.

▲ Ask the children to name foods that they eat frozen, foods that they eat hot and foods that are neither hot nor frozen when eaten.

▲ Write their answers on the chart.

▲ Compare the three lists.

▲ What does each list have in common? How are the three lists different from each other?

ice cream popsicles	crackers apples pretzels	Hot chocolate chicken soup meatloaf
Frozen	Not Frozen Not Hot	Hot

▲ Interest area
Group time

Science & math principle

Teaches about diversity as the children generate the list of foods.

Science & math skill

Encourages communication skills.

Science & math attitude

Develops a desire for knowledge (curiosity).

▼

You need

chart paper and marker

Questions you might ask

▲ Which list has the most items in it? Why?

▲ Is there a food that could go in all three categories of the list?

Extension

Taste some of the foods from the three categories on the list.

More challenging for older children

Place the list in the Science and Math Center so they can add to it.

Modifications for younger children

Do this activity at four different group times. The first time, list foods that are frozen; the second time, list foods that are hot; the third time, list foods that we eat that are neither cold nor hot; and the fourth time, talk about all three lists.

Science & math principle

Teaches children to observe the interaction between the scoop, the cups and the ice.

Science & math skill

Develops observation skills.

Science & math attitude

Encourages children's curiosity.

▼

You need

enough bags of crushed ice to half fill a large tub or water table

several ice cream scoops (use different kinds of scoops to make this activity more interesting)

clear plastic containers, different sizes

Ice Scooping

Activity

▲ Place the ice in the water table (or tub).
▲ Give the children scoops to fill the different containers.
▲ The children scoop, fill and dump while they count how many scoops it takes to fill the different containers.

Questions you might ask

▲ How many scoops did it take to fill this container? and this one?
▲ Did it take the same number of scoops?

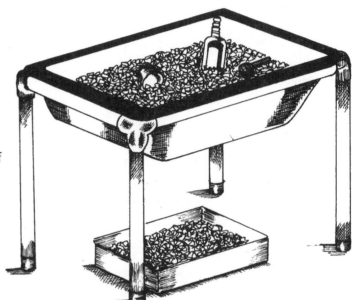

Extension

Make a list of how many scoops it took to fill different containers (for example, bowl one—3 scoops, bowl two—5 scoops, bowl three—2 scoops, and so on). Discuss why there is a difference.

Note: Make sure the bowls or containers are significantly different in size.

More challenging for older children

As the ice melts, they can compare how many scoops of water it took to fill the various containers as compared to how many scoops of ice.

Modifications for younger children

Give them mittens to wear as they will likely play with the ice as well as scoop it.

Ice Cube Sorting

Activity

▲ Place the dog dish, bowl of ice and tongs inside the dishpan or in a water table.

▲ Place a small towel close by to keep things dry.

▲ Using the tongs, children move the blocks of ice from the container to the dog dish. If they are large, they go in one side of the dish; if they are small, they go in the other.

▲ Pour off the water from the dishpan and from the dog dish frequently.

Extension

Freeze objects inside the ice cubes like twigs in one set and leaves in the other. The children sort them: cubes with twigs in one side, leaves in the other, or freeze two colors of ice cubes and have the children sort them by color.

More challenging for older children

Record how many large pieces and how many small pieces of ice were moved to the dog dish.

Modifications for younger children

If they cannot use tongs successfully, let them pick up the ice with their hands. If they don't like the coldness, put gloves or mittens on their hands.

Interest area
Sand and water

Science & math principle
Teaches children to notice the patterns in the ice.

Science & math skill
Teaches children to organize the patterns by sorting them.

Science & math attitude
Develops children's patience.

▼

You need

plastic dish with 2 sections (I use a double-sided dog dish)
small towel
large and small blocks of ice in a container
tongs
dishpan or water table

Questions you might ask

▲ How are the pieces of ice the same? different?

▲ Where might you find ice?

▲ How is ice useful?

Interest area
Sand and water

Science & math principle

Teaches about cause and effect when they test inclines and materials.

Science & math skill

Develops observation skills.

Science & math attitude

Encourages children to keep an open mind about the results.

You need

ice chest filled with ice cubes
metal cookie baking sheet, without sides
wooden board
cardboard gift-wrapping tube
large PVC pipe at least 3″ (8 cm) in diameter
sheet of Plexiglas
large tub or water table
basket

Questions you might ask

▲ As the ice melted, did it slide faster or slower? Did it travel further?
▲ What would happen if ice had feet?

Ice Race

Activity

▲ Place all the ice in the water table (or tub).
▲ Put the wooden board, cardboard tube, PVC pipe and sheet of Plexiglas in or near the basket and place the basket beside the table.
▲ The children experiment with each equipment piece to see which surface allows them to slide the ice the farthest and the fastest.

Note: This is an open-ended exploration activity. There is no correct way to do the activity. The children may discover that if they hold the equipment on an incline, the ice travels faster and farther; or that different equipment (surfaces) may allow the ice to travel faster (for example, ice will travel faster on Plexiglas than on cardboard). Also, the children may realize that the ice travels faster as it melts.

Extension

Make the ice cubes different sizes. The children can measure the distances the ice slides and compare size and distance relationships (for example, larger cubes weigh more and slide farther).

More challenging for older children

Encourage the children to create a raceway in the tub. They can draw a start and a finish line with permanent markers. (Remove the start and the finish lines with fingernail polish remover when the experiment is over.)

Modifications for younger children

Give the children lots of time to explore the ice.

Measuring the Temperature

Activity

▲ Before the children arrive, fill one jar three-fourths full of ice and the other jar three-fourths full of warm water.

Note: Use a small electric coffee cup warming tray to keep the water in the jar warm.

▲ Place the thermometer, the jar of ice and the jar of hot water in the tray.

▲ The children move the thermometer from one jar to the other, observing the temperature change from hot to cold and from cold to hot.

Questions you might ask

▲ What happens when you move the thermometer from the cold to the hot? from hot to cold?

▲ What do the numbers tell you about the water and the ice? Could you tell if it was hot or cold just by looking at the numbers on the thermometer?

Extension

Make ice cubes from sand and water, sugar and water, and vinegar and water. Test them to see if they are colder or warmer than frozen water.

More challenging for older children

The children can record the changes in the temperature as the ice melts and the warm water cools.

Modifications for younger children

Do this in supervised small groups; do not leave this activity in a center for free exploration.

Interest area
Science and math

Science & math principle
Teaches about the properties of warm and cold water.

Science & math skills
Teaches children to compare information.

Science & math attitude
Helps children become comfortable with science and math.

▼

You need

1 child-proof thermometer (If you cannot find a child-proof, mercury thermometer get a metal candy thermometer.)

2 clear plastic jars about 6″ (15 cm) tall (e.g., mayonnaise jars)

small ice chest filled with ice

warm water

tray with 3″ (8 cm) sides

Science & math principle

Teaches about using a model as children work with pretend ice.

Science & math skill

Develops communication skills.

Science & math attitude

Helps children develop patience as they wait for the pretend ice to dry.

▼

You need

1 cup (250 mL) of Epsom salts
water
spoon
pan
2 glass jars with lids
stove or hot plate
2 dark colored trays (like cafeteria trays) with 1" (2.5 cm) sides
number line
chart table and marker

Pretend Ice

Activity

▲ Mix Epsom salts and ¾ cup (175 ml) of water in a pan. Bring to a boil on medium high until the mixture is clear.

▲ When the mixture cools to room temperature, pour it into a jar and screw on the lid tightly. Fill a second jar with ¾ cup (175 ml) of plain water.

▲ Ask the children to describe the liquids in the jars. Ask why they think both jars have water in them.

▲ Pour the Epsom salts mixture onto the dark tray (as a contrasting medium), covering the bottom about ½ inch (13 mm).

▲ Ask the children to predict what will happen when you put the tray aside in a cabinet or closet. List all of their guesses on the chart tablet.

▲ Pour the plain water in the second tray and ask the children to predict what will happen when you put this tray in the freezer. List all their guesses.

▲ Wait two weeks by marking off the days on a number line. When the two weeks have passed, remove the ice tray. Examine it carefully, checking the children's predictions and describing the characteristics of the ice.

▲ Next, retrieve the tray of Epsom salts. It will look like a "lake" of cracked "ice" but it will not feel cold. Have the children compare the real ice to the "pretend" ice made with Epsom salts. Show them all the ingredients and talk about how they think the pretend ice formed.

Questions you might ask

▲ How are the ice and pretend ice the same? different?

▲ Which ice do you think you could hold for the longest time?

Extension

Place large sheets of black paper at the easel and allow the children to paint with the leftover Epsom salts mixture. It will apply to the paper like water, but it will dry and look icy.

More challenging for older children

Encourage the children to brainstorm other ways that ice can be represented. Some solutions are Milar, thin metal sheeting, silver fabric or aluminum foil.

Modifications for younger children

Talk about their predictions rather than recording them. Supervise them well (we do not want them drinking Epsom salts!).

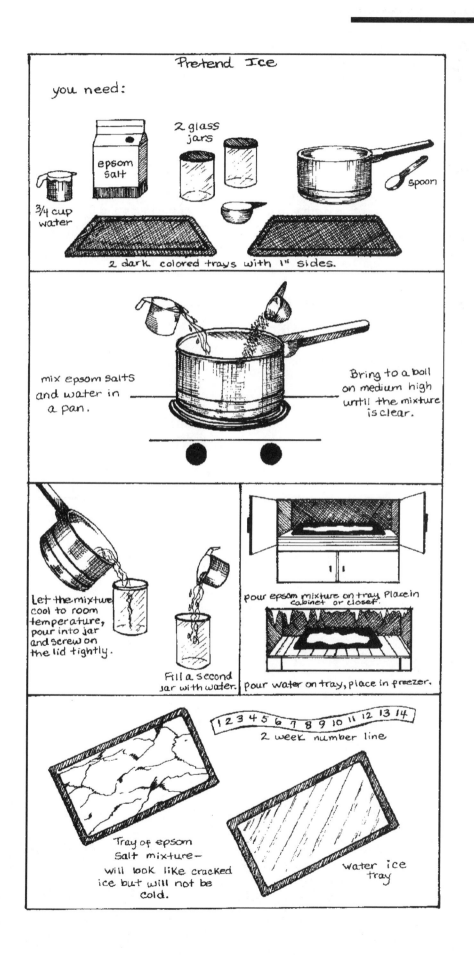

Pretend Ice

you need:

3/4 cup water

epsom salt

2 glass jars

spoon

2 dark colored trays with 1" sides.

mix epsom salts and water in a pan.

Bring to a boil on medium high until the mixture is clear.

Let the mixture cool to room temperature, pour into jar and screw on the lid tightly.

Fill a second jar with water.

pour epsom mixture on tray. Place in cabinet or closet.

pour water on tray, place in freezer.

1 2 3 4 5 6 7 8 9 10 11 12 13 14
2 week number line

Tray of epsom salt mixture—
will look like cracked ice but will not be cold.

water ice tray

Ice Melt

Science & math principle

Teaches about interaction when comparing the covered and uncovered mug.

Science & math skill

Teaches children to compare information.

Science & math attitude

Helps children develop patience as they wait for melting.

▼

You need

2 coffee mugs
enough ice cubes to fill both mugs
sheet of cotton batting
scissors
clear cellophane
tray
tape
chart tablet
marker

Activity

▲ Make a prediction chart.
▲ Fill both mugs with ice.
▲ Cover the mouth of both mugs with clear cellophane.
▲ Cut a piece of cotton batting to fit around one of the mugs and tape it in place.
▲ Ask the children to predict how long it will take for the ice to melt in the cotton-covered mug and how long in the other mug.
▲ Check the actual melting time for each and compare the results.

Questions you might ask

▲ Which mug of ice melted first? Why?
▲ Why do we wear coats in the winter?

Extension

Wrap the mugs with different materials. Chart the melting process.

More challenging for older children

Put the children in charge of keeping track of the time with a timer or a clock, if that is appropriate.

Modifications for younger children

Check the melting process every forty-five minutes so the children will be able to see a greater change.

Ice Color Mixing

Activity

▲ The children make one block of ice in two colors, yellow and blue.
▲ First, fill the milk carton one-half full of water.
▲ Add yellow food coloring, enough to color the water to a deep yellow (use cake decorating paste to get a deep, rich color).
▲ Freeze it solid.
▲ Remove the milk carton from the freezer and fill the remaining half with water colored light blue.
▲ Refreeze the milk carton. With this block of ice, do the following study using the scientific method.

State the Question

What will happen to the two colors when the block of ice melts?

State the Hypothesis (as a class or individually)

We think the block of ice will _____ , and _____ will happen.

Note: The children's statements will vary considerably, depending on the children's knowledge and the diversity of their experiences. Some responses I have gotten over the years are: "Water comes...all over the floor"; "a mess"; "ice"; "yellow water"; and some have predicted that the melted water would be green. Be sure to write each child's response on a chart. They can check the results against their predictions when the project is completed.

Method of Research

We will peel the paper from the carton and put the block of ice in the dishpan. We will place it in a warm place and check it frequently to see what happens.

Check the Hypothesis

When the ice melts, we will see what happened and check our hypothesis.

Results

The ice melted and the yellow and blue water combined and became green.

Extension

Try the activity with other colors.

More challenging for older children

Ask them to predict how long it will take for the ice to melt. Have them measure the time it takes for the ice to melt, as well as what happens when it melts.

Modifications for younger children

Speed up the process by freezing a pint of water in two halves, just as described above. Make sure the children don't try to taste the melting ice.

Interest area
Science and math

Science & math principle
Teaches about change as the colors blend.

Science & math skill
Develops observation skills.

Science & math attitude
Encourages children to keep an open mind when they check their predictions.

▼

You need

1 half-gallon (2-liter) paper milk carton
yellow and blue food coloring
water
dishpan
chart tablet and marker

Questions you might ask

▲ What would happen if the colors were red and yellow?
▲ Why do you think the ice melted?

Peanut Shell Print

Science & math principle

Teaches cause and effect as children print with the peanut shells.

Science & math skill

Develops observation skills.

Science & math attitude

Develops children's curiosity.

You need

peanut shell halves
3 colors of liquid tempera paint
3 paintbrushes
paper (any size)
1 divided tray with 3 compartments, or 3 small containers
newspaper

Activity

▲ Cover the work area with newspaper.
▲ Pour the three colors of paint into the three compartments or containers.
▲ The children paint a rounded half of the peanut shell any color. They place the paper over it and gently press down to make a print.
▲ Encourage the children to do this repeatedly, changing colors each time. This will create a peanut shell design.
▲ They can also try pressing the peanut shell onto the paper. Compare the two methods.

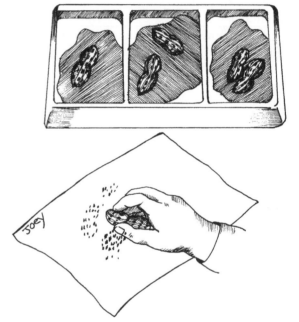

Questions you might ask

▲ What happened when your shell prints over-lapped?
▲ What would happen if you paint half the shell one color and the other half another, and then you printed with it?

Extension

Encourage the children to cut out as many peanut shell prints as they like and make a class collage of their peanut impressions.

More challenging for older children

Encourage them to try different ways to create a peanut print on their paper (painting the edges of the peanut shell and imprinting peanut "footprints" across their paper or dipping the end of the peanut and printing with it).

Modifications for younger children

It will be difficult for younger children to hold the peanut while they are painting it without covering their hands in paint. Stabilize the shells for painting by placing them in an empty egg carton.

Peanut Measuring

Activity

▲ Make peanut measurers (a nonstandard measuring device) by cutting out peanut shapes from brown construction paper. Make these peanut shapes 2″ (5 cm) long.

▲ Laminate or cover them with clear contact paper.

▲ As the children build in the Block Center, have them use the peanut shapes to measure their constructions. Ask them how long, how wide and how tall their construction is in terms of "peanuts" (children should lay these shapes end to end along object they are measuring).

Build a structure.

measure the length.

How long is it?

More challenging for older children

Encourage the children to measure objects with their peanut measurers and then compare the results with those of a standard measuring tool like a tape measure.

Modifications for younger children

Demonstrate measuring with the peanut measurer. If the children are using bigger blocks, make the peanut measurers as long as the blocks or make more measurers.

Interest area
Blocks

Science & math principle
Teaches about measuring (scale).

Science & math skill
Encourages children to compare the different lengths.

Science & math attitude
Teaches children confidence in their skills.

▼

You need

brown construction paper
black marker
scissors
clear contact paper
blocks

Questions you might ask

▲ What other tools could you use to measure the road?

▲ What can you build that will be five peanuts wide?

Extension

Encourage the children to make their own nonstandard measuring tools to place in the Block Center.

Interest area
Group time

Science & math principle
Teaches about change as the peanuts become peanut butter.

Science & math skill
Develops observation skills.

Science & math attitude
Helps children become comfortable with the tools used in science and math.

▼

You need

shelled, roasted peanuts (if you want the children to roast their own, purchase raw peanuts)
bowl
food processor or blender
measuring cups and spoons
vegetable oil
salt
spatula
crackers
knife
paper plates and napkins

Peanut Butter

Activity

▲ Read or tell the story below. Afterward, make peanut butter as a class. If you want to start from scratch and use raw peanuts, bake them in the oven at 300º F (150º C) for 45 minutes. Set them aside to cool and have the children help shell them. If you don't want to start from scratch, just purchase a large jar of preshelled, roasted peanuts and use the recipe as your starting point for this activity.

Note: In my own classes, we usually roast a few of our own, shell them and add them to the recipe along with the preshelled, roasted peanuts from the jar.

Peanut Butter Recipe

Pour 1 cup (250 ml) of unshelled peanuts into the food processor. Add 2 tablespoons (30 ml) of vegetable oil and ¼ teaspoonful (1.5 grams) of salt. Blend until creamy. Serve on crackers, like in the story below.

PEANUT, PEANUT BUTTER

by Sharon MacDonald

Meg found a peanut on the ground. She rolled it around in her hands. She shook it close to her ear. She liked the sound it made.

She showed the peanut to her Grandma, who was sitting on the front porch with Grandpa.

"Look what I found," Meg said, shaking the peanut again. "Hear the sound?"

Grandma smiled. "Yes, I do."

"What else can I do with my peanut?" Meg asked.

Grandma thought for a minute. "When I was a little girl," she said, "my family went to a picnic. It was summer. We had peanut races. We had to roll the peanut along the ground with our noses! You could try that."

Meg got down and tried to roll the peanut along the ground with her nose, but her knees hurt. She did not think it was much fun.

Meg stood up. "Grandpa," she asked, "what can I do with my peanut?"

Grandpa leaned back in his chair and thought for a while. "When I was a little boy," he answered, "we went to a family get-together in the spring. We played games and races, too. I had to hold a peanut under my chin and run as fast as I could to the finish line without dropping it."

Meg tried that, but she could not run very far. The peanut kept falling out. She didn't think that was much fun, either.

Meg showed the peanut to Mom and told her what Grandma and Grandpa had done with a peanut when they were young.

Mom took the peanut and shook it, too. "Let's see," she said thoughtfully. "There are other things we could do with the peanut."

"What?" Meg asked.

"When I was a little girl, my Aunt Frances made peanut butter from peanuts. It tasted so good!" said Mom.

Meg said, "Oh please, Mom, let's make peanut butter with my peanut!"

Meg's mother shook her head. "One peanut is not enough, Meg. It takes lots of peanuts to make peanut butter. We could buy more at the store. Let's go and buy a whole bag of them."

And they did.

When they got home, Meg and Mom shelled the peanuts. Mom got out the food processor and some vegetable oil. Meg put all the shelled peanuts in the food processor and Mom added a little oil. Meg turned on the food processor at the highest speed. It made a lot of noise. After a few minutes, Meg turned off the food processor and removed the top. When she looked inside, she saw smooth peanut butter! It smelled very good.

"Mom," Meg said excitedly, "it looks just like peanut butter we buy in the jar!"

Meg and her Mom got some crackers from the cabinet. Meg spread the peanut butter on four crackers and put them on a plate. They took the plate outside to the porch and shared the peanut butter with Grandpa and Grandma. They decided it was the best peanut butter they had ever tasted. Meg was very glad she had found a peanut on the ground.

Questions you might ask

▲ Why do you think Meg needed more than just one peanut to make peanut butter?
▲ What else could Meg do with her peanut?

More challenging for older children

Brainstorm other ways to make peanut butter.

Modifications for younger children

Do this at two different times. Tell the story first, then make the peanut butter. Involve the children as much as possible and supervise carefully. You will want to stop occasionally for the children to taste the peanut butter they are making.

Extension

Make other types of butter like apple butter, nutmeg butter, cocoa butter and coconut butter.

▼ 147

Peanut Game

Interest area
Group time

Science & math principle

Teaches about a system as they follow the steps in the game.

Science & math skill

Develops observation skills.

Science & math attitude

Teaches children to cooperate as they play the game.

▼

You need

one peanut in the shell

Questions you might ask

▲ How do you keep people from knowing when you have the peanut?

▲ What other object could you hide in this game?

Extension

Brainstorm ways to do this activity with objects too big to hide in hands.

Activity

▲ The purpose of this activity is to have the children guess who's got the peanut.
▲ Sit in a circle with hands together in front.
▲ Give the peanut to one of the children.
▲ That child goes around the circle and slides his hands through each of the other children's hands.
▲ He secretly passes off the peanut to one of the children in the group. To keep from giving away who received the peanut, he continues until he has touched the hands of all the children. As he goes around the circle, say the following chant:

PEANUT, PEANUT, WHERE CAN YOU BE?

by Sharon MacDonald

Peanut, peanut
Where can you be?
Peanut, peanut,
1...2...3...!

▲ All of the children keep their hands closed even if they don't have the peanut.
▲ Then, starting with the person sitting to the right of the peanut passer, each child gets to guess who has the peanut (one guess each).
▲ The guesses continue until the peanut is found. The child with the peanut then says the following chant:

I've got the peanut.
Look and see.
I'll pass the peanut.
1...2...3...!

▲ Next, the child who guessed correctly takes a turn around the circle. Continue the game until each child has had a turn.

More challenging for older children

Have the children put their hands together behind them so there will be no visual clues when the peanut is given away.

Modifications for younger children

Do this in groups of three or four because it takes too long in a larger group for each child to have a turn. You may have to give the children a sock to catch the peanut (it is difficult for most younger children to remember to keep their hands together and they easily give themselves away).

Peanut Predicting

Activity

▲ Fill the small jar with peanuts in the shell and secure the lid.

▲ Place the jar on the small tray.

▲ Write the numbers one through fifteen on the adding-machine tape.

▲ Pass the jar round and let the children guess how many peanuts there are in the jar.

▲ Write each guess on a chart.

▲ When the children have guessed, ask them to explain how they could check if their guesses were correct (they will verify by counting).

▲ As you count the peanuts, place one peanut on each number on the number line (this will help them associate the written number with the spoken one as they count, and to learn about one-to-one correspondence).

▲ Check their original guesses on the chart.

▲ Place the jar on the tray with the number line so the children can recount the peanuts in the Science and Math Center.

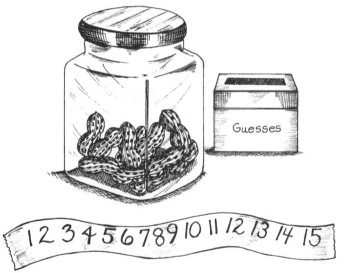

Extension

Place the peanuts in an opaque jar. Have them shake the jar and predict the number of peanuts inside based on sound alone.

More challenging for older children

Put thirty to forty peanuts in the jar so that when the children are counting the peanuts, they can group by tens.

Modifications for younger children

Place four or five peanuts in the jar and count them with the children. Most younger children are not developmentally ready to understand and use the number line.

Interest area
Group time

Science & math principle

Teaches about population as they try to guess the number of peanuts in the jar.

Science & math skill

Encourages children to compare their estimates to the actual number.

Science & math attitude

Encourages children to keep an open mind and to change their minds.

▼

You need

small plastic jar with a lid
10 to 15 peanuts in the shell
tray
adding-machine tape
  number line
chart tablet and marker

Questions you might ask

▲ Why did you guess that there were ____ peanuts?

▲ If you could try again, how many would you guess?

Peanut Butter & Apple Slices

Interest area
Group time

Science & math principle
Teaches about the interaction of peanut butter on wet and dry apple slices.

Science & math skill
Encourages children to compare the results using wet and dry apple slices.

Science & math attitude
Helps children develop patience.

▼

You need for each child

2 apple slices
2 small containers of peanut butter (medicine cups work well)
2 plastic knives
2 small paper plates
napkins

Activity

▲ Before the children arrive, prepare two sets of plates for each child. On each plate put one small container of peanut butter, one knife, one apple slice and a napkin.
▲ Pass out one plate to each child.
▲ Talk with the children about the qualities of peanut butter.
▲ Encourage each child to take a small taste of peanut butter.
▲ Talk about how sticky it is in their mouths.
▲ After they all agree that peanut butter is really sticky, ask them to try to smear the peanut butter on their apple slice (it will slip and slide all over the apple).
▲ Some of the children may be able to balance a lump of it on the apple slice, but most of it will end up on the plate.
▲ Ask them to taste their apple and peanut butter snack.
▲ While they eat, talk about why it was so hard to get something so sticky like peanut butter to spread on the apple and stay there.
▲ After they have thrown away their plates and other items, pass out the second plate with the apple slice, the container of peanut butter, the plastic knife and a napkin.
▲ Encourage the children to dry their apple slice carefully with the napkin before they spread on the peanut butter.
▲ Talk about what happened and how this was different from the first time they tried to spread peanut butter on the apple slice.

Questions you might ask

▲ Why do you think it was so hard to put peanut butter on the apple slice the first time?
▲ On which apple slice were you able to get the most peanut butter? Why?

Extension

Compare spreading peanut butter on apple slices to spreading it on other food like bread, crackers, rice cakes and celery.

More challenging for older children

Brainstorm ways to get the peanut butter to stick to the apple slice and then try some of their suggestions.

Modifications for younger children

Expect most of them to eat the apple with little or no peanut butter on it, lick the knife and eat the peanut butter out of the container. It will be difficult for them to hold the apple slice and spread the peanut butter at the same time, so give the children a toothpick to steady the apple slice while they try to spread on the peanut butter.

Peanut Puzzle

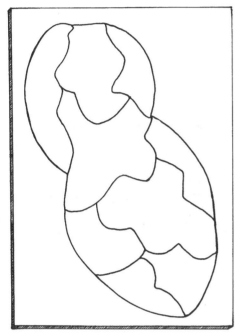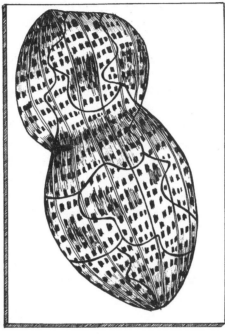

Interest area
Manipulatives

Science & math principle
Teaches about interaction as they fit the pieces together.

Science & math skill
Develops observation skills.

Science & math attitude
Teaches children to be curious.

▼

You need

photograph or drawing of a peanut (or peanuts in the shell)
scissors
rubber cement
tagboard sheets
marker or pencil
resealable plastic bag to hold the puzzle pieces
clear contact paper

Activity

▲ Cut the tag board sheets in half.
▲ Make a peanut puzzle by gluing the peanut picture to half of the tagboard sheet using the rubber cement.
▲ Cut the picture into puzzle pieces (you can decide how many pieces based on your children's ability level).
▲ Trace the individual puzzle pieces on the other half of the tag board sheet to form a base on which the puzzle can be put together.
▲ Put the puzzle pieces and base in a plastic bag and place in the Manipulatives Center.
▲ The children put the pieces of the peanut puzzle together on the base.

Extension

Divide the class into small groups and have each group draw a large picture of a peanut, about 18 x 24 inches (45 cm x 60 cm). Color the puzzle, then cut it into several puzzle pieces. The groups exchange puzzles and try to solve them.

More challenging for older children

Cut the puzzle into a larger number of pieces and do not provide the base.

Modifications for younger children

Cut the puzzle into a few large pieces.

Questions you might ask

▲ How are the pieces of the puzzle alike? different?
▲ What do you think makes something a puzzle?

.

peanuts

Interest area
Science & math

Science & math principle
Teaches children to find a common pattern in a group of peanuts.

Science & math skill
Teaches children to organize information.

Science & math attitude
Encourages children to work together to solve a problem (cooperation).

▼

You need

bag of cooked peanuts in shells
large sheets of paper (1 per 4 children)
marker

Questions you might ask

▲ How many peanuts are in each group?
▲ Why was it hard to divide them?

Extension

Encourage each group to make a picture representation of their groupings.

Peanut Grouping

Activity

▲ Draw a Venn diagram (2 overlapping circles) on large sheets of paper, one sheet for every four children.
▲ Divide the children into teams of four.
▲ Give each team ten peanuts and a Venn diagram.
▲ Explain to the children that they are going to decide together (as teams) how to group the peanuts on the Venn diagram, but that they do not have to have equal numbers of peanuts in each group.
▲ Have them sort peanuts into three groups by placing them on the Venn diagram. In one circle, the peanuts all have a common characteristic; in the other circle, they have a common characteristic that is different from the characteristic in the first circle; and in the area where the two circles overlap, all the peanuts have both characteristics in common.
▲ Since peanuts are very similar, this activity will involve a lot of negotiating and compromising on the part of individual team members.
Note: Save all the peanuts from all the activities and use them to make peanut butter in Peanut Butter, on page 146.

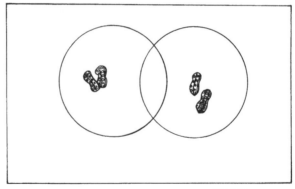

More challenging for older children

Ask the children to generate a list of the characteristics they used to group the peanuts. Have them compare their list to the other teams' lists.

Modifications for younger children

Let the children explore, crack and eat some peanuts before you try this activity. Ask the children to sort the peanuts into two groups.

Peanut Weighing

Interest area
Science and math

Science & math
principle
Teaches children about scale
as they measure peanuts.

Science & math
skill
Teaches children to compare
the weights.

Science & math
attitude
Encourages a desire for
knowledge (curiosity).

You need

balance scale
container full of peanuts in
the shell

Activity

▲ The children weigh the peanuts.
▲ They fill one side of the balance scale with peanuts and experiment with how many it will take on the other side to balance the scale.
▲ They can make an approximation by placing precounted groups of peanuts on the other side or by counting them one at a time as they place them on the other side.
▲ Ask them why they think ten peanuts on one side of the scale may not necessarily balance with ten peanuts on the other side.

Questions you might ask

▲ Why do you think the peanut makes the scale drop on one side?
▲ What would happen if we filled one side with rocks and the other with peanuts?

Extension

Give the children a numerical kitchen scale to weigh different quantities of peanuts.

More challenging for older children

Give them metal weights to use on one side and see how many peanuts it takes to balance specific amounts of weight on the other.

Modifications for younger children

Demonstrate how to use the scale in small groups. Give them a limited number of peanuts to balance. Don't be surprised if many of the peanuts are eaten!

Peanut Dropping

Science & math principle

Teaches about interaction as children use their hands and the tongs to move the peanuts.

Science & math skill

Develops observation skills.

Science & math attitude

Encourages children to be curious.

▼

You need

large clear jar (like a pickle jar)
small clear jar (like an olive jar)
Fun Tac (a brand name adhesive)
tongs
small bowl
12 to 15 peanuts in the shell
large tray

Activity

▲ Before the children arrive, place a large ball of *Fun Tac* on the bottom (inside) of the large jar.
▲ Place the small jar on top of the sticky ball of *Fun Tac* so it is centered and secured inside the larger jar.
▲ Place the tongs, the bowl with peanuts and the large jar with the small jar inside on the tray.
▲ Place the tray on the floor in the Science and Math Center (in a low traffic area).
▲ Using the tongs, the children pick up the peanuts one at a time from the bowl and try dropping them into the small jar while standing over it. Some peanuts will end up inside the larger jar and some on the floor.
Note: Dump the jars to retrieve the peanuts or have the children pick them out with the tongs.

Questions you might ask

▲ Why is it harder to get the peanut into the smaller jar than the larger one?
▲ How can you find out how many peanuts you got in the smaller jar?

Extension

Use progressively smaller jars inside the larger one to make the peanut drop more difficult (they will have more success as their fine motor skills improve). Have them discuss which is easier and which is harder, and why.

More challenging for older children

Have children work in pairs. Challenge them to figure out a way to keep score.

Modifications for younger children

Children sit, rather than stand, over the jar. Many will not be able to manipulate the tongs, so let them use their hands. Later, try it again with the tongs.

Peanuts vs. Peanut Butter

Activity

▲ For each child, place a container of peanut butter, a spoon and four peanuts on a plate. Pass out the plates.

▲ Ask the children to predict whether the peanut and the peanut butter will be: salty, smooth, crunchy, sticky and hard. Have them try out (verify) their predictions by tasting the peanut butter and peanuts. Vote on the characteristics they have been asked to predict.

▲ Make a chart to record the class predictions.

State the Problem or Question
Which will have the following characteristics, the peanut or the peanut butter: salty, smooth, crunchy, sticky and hard?

State the Hypothesis
We believe that the peanut will have these characteristics: _____, _____, _____, and the peanut butter will have these characteristics: _____, _____, _____.

Method of Research
We will taste each one and decide.

Checking the Hypothesis
We will look at our chart to verify if our predictions were correct.

Results
We found that the peanut was _____, and the peanut butter was _____.

Questions you might ask

▲ Which do you like the best? Why?

▲ Do you think it is possible for peanuts and peanut butter to have the same characteristic? Which one?

Extension

Encourage the children to think up other words that describe peanuts and peanut butter in relationship to how each looks.

More challenging for older children

Ask them to write down (or write for them) their individual predictions and check them.

Modifications for younger children

Limit the number of characteristics to be predicted to two.

Interest area
Science and math

Science & math principle
Teaches about properties of peanuts and peanut butter.

Science & math skill
Develops observation skills.

Science & math attitude
Helps children be comfortable with science and math.

▼

You need

peanut butter in small containers, with spoons
4 shelled peanuts per child
1 paper plate per child
chart tablet and marker

One-Third of a Hot Potato

Interest area
Cooking and snack

Science & math principle
Teaches children about how potatoes change when cooked.

Science & math skill
Develops observation skills.

Science & math attitude
Encourages children's curiosity.

▼

You need

small or medium-size Idaho potatoes (sufficient for each child to have about 1/3 of a potato)
vegetable brush
margarine
forks
stove
paper plates
salt (optional)

Activity

▲ The children bake the whole potatoes (with the skins) following the rebus directions and taste the results.
▲ Preheat the oven to 425° F (220° C).
▲ The children wash, scrub and dry the potatoes, then put margarine on their hands and rub it on each potato.
▲ Place the potatoes on the oven rack and bake for 40 to 60 minutes (depending on the size of the potatoes). When half the time has elapsed, pull out the rack and quickly and carefully puncture the skin of each potato with a fork (this allows steam to escape). Continue baking.
▲ When they are done, allow them to cool. Cut each potato into 1/3 sections. Do not remove the skin (a great source of vitamins and minerals!). While you are cutting and serving, talk about part-and-whole relationships.
▲ Serve on the plates. The children salt or put more margarine on their potato if they want. Talk about how the potato changed and how it tastes.

Questions you might ask

▲ What is the best thing about this potato?
▲ Have you ever tasted anything like it before?

Extension

Bake potatoes and serve them with different toppings such as bacon bits, cheese, sour cream, chives, parsley or deviled ham.

More challenging for older children

Ask the children to retell the baking process by drawing it as a picture sequence.

Modifications for younger children

The whole process from preparing to eating takes a long time. Set this up so that the children can be involved in each step but don't have to work too long at one time.

Share with 2 friends.

Wash.

cover with margarine.

Bake.

Puncture and finish
baking.

Cool.

cut into 3 parts.

How Ten Became Mashed

Interest area
Cooking and snack

Science & math principle
Teaches children how potatoes change when cooked.

Science & math skill
Develops observation skills.

Science & math attitude
Encourages children's curiosity.

▼

You need

10 potatoes
vegetable brush
potato peelers
chopping block
knife
water
salt
large cooking pot
stove
potato masher
margarine
salt (optional)
small paper bowls or plates
plastic spoons
napkins

Activity

▲ The children "read" the picture directions and follow the steps to make mashed potatoes.

▲ The children wash, scrub and dry the potatoes.

▲ Peel the potatoes and cut into 4" or 5" (10 cm or 12 cm) cubes. If the children do the cutting, be sure it is closely supervised.

▲ Place the cubes in the pot and cover with water. Bring to a rolling boil and maintain it for 20 to 40 minutes (they are done when you can easily push a fork into any of the cubes).

▲ The adult drains the pot.

▲ Add a tablespoon of margarine, then mash the potatoes with the potato masher (taking turns) until smooth. Add salt if the children choose.

▲ Serve a teaspoon or so to each child. Have them taste the potatoes if they want, but please do not insist that they do so. Talk about the change in the potatoes and how the change came about.

wash

cut

cook

Drain

Add one tablespoon margarine.

mash.

Serve.

Eat!

Questions you might ask

▲ How did the potatoes change?
▲ What else could you use to mash potatoes?

Extension

Make mashed potato balls. Add two well-beaten eggs and a little parsley, and roll the mashed potatoes into balls. Bake them in a lightly greased muffin tin until crisp. Turn the balls in the muffin tin to brown the bottom. Let them cool and serve them for snack!

More challenging for older children

Check the potatoes every ten minutes by removing one of the potato cubes and cutting it. Examine the cube to see what is happening. Test for hardness and changes in texture.

Modifications for younger children

In this activity there is more for the adult to do than the children. This being the case, allow them lots of time to wash and scrub the potatoes and to mash them. I recommend that the adult does the cutting. Make sure the children practice safety around the boiling water.

You need

French fries (purchase enough fries from a fast food outlet for each child to have at least one)
4 large potatoes
vegetable oil
¼ cup (60 ml) measure
9" x 13" (22 cm x 32 cm) baking pan
salt
paper towels
paper plates
potato cutter
bowl half full of very cold water
chart tablet and marker
oven

Oven Fries

Activity

▲ The children "read" and follow the picture directions to make the French fries. (Set aside the precooked French fries for later use.)
▲ Preheat the oven to 450° F (230° C).
▲ The children wash, scrub and dry the potatoes.
▲ Use the potato cutter to cut the potatoes into small strips. Usually the cutter comes in ½" (13 mm) squares. If you do not have a cutter, cut the potatoes into ½" (13 mm) strips. As you cut the strips, place them in the cold water in the bowl. All the strips need a cold water bath.
▲ Remove and dry them well on the paper toweling.
▲ Dry the bowl and add the potatoes. Pour ¼ cup (60 ml) oil over the potatoes. Make sure they are well-coated with vegetable oil.
▲ Spread the potatoes onto the baking pan. Cook the potatoes in the oven for 30 to 40 minutes. Turn them several times while browning.
▲ Drain them on paper towels. Sprinkle with salt.
▲ Serve the oven-baked and the purchased French fried potatoes to the children on the paper plates. After the children have tasted both, talk about the similarities and differences and make a list. Encourage the children to add to the list.

Questions you might ask

▲ When you look at the French fries we bought and the French fries we made, how are the potatoes alike? different?
▲ What other tool could you use to cut the potato?

Extension

Ask the children to bring in recipes for different ways to cook French fries.

More challenging for older children

Cook deep-fried French fries on one day and oven-baked fries on another. Have them compare the taste of each and talk about the differences in cooking processes.

Modifications for younger children

It may be difficult for younger children to cut the potatoes into strips or to use the potato-cutter. Involve them in the rest of the activity, making sure they are careful around the hot stove and hot potatoes.

Cut the potato
with a potato cutter.

cold water

Place in cold water.

Dry the strips and the bowl.

cover the sticks with oil.

Spread on baking pan.

Bake and turn.

Drain and cool.

Serve and eat.

Sheer Potatoes

Interest area
Dramatic play

Science & math principle
Teaches children about using a model in place of a real object.

Science & math skill
Encourages children to communicate information they learned about potatoes.

Science & math attitude
Develops children's confidence.

▼

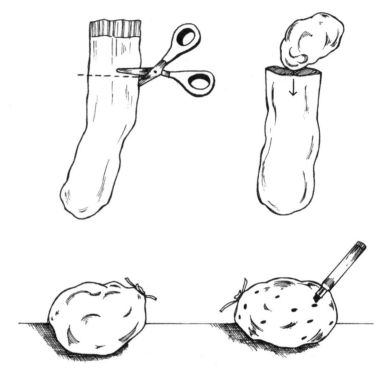

You need

clean brown panty hose, 1 per child, or knee-high stockings, 5 per child
twist ties
black markers
scissors

Questions you might ask

▲ What makes your potato look like a real potato?
▲ When would you use a pretend potato instead of a real potato?

Extension

Brainstorm what other pretend foods they can make with stockings.

Activity

▲ The children make stocking "potatoes" by cutting off the foot of the stocking about up to the ankle.
▲ Use the rest of the stocking to stuff into the foot part.
▲ Fill and shape it until it looks like a potato.
▲ Tie off the closure with a twist tie or a small knot.
▲ The children decide how many "eyes" they want on their potatoes, then make black dots on the stockings with a black marker.
▲ Use the stocking potatoes to play games like Hot Potato, Potato Racing and Rolling the Potato, or "cook" with it in the Dramatic Play Center.
Note: If you use permanent markers to draw the "eyes" on the potatoes, you can wash them. If you plan on washing them, tie a knot instead of using a twist tie.

More challenging for older children

Encourage the children to invent a game to use with their potato.

Modifications for younger children

Talk about the differences between things that are real and things that are pretend, like the stocking potato. You will have to help the children use the twist tie or tie the knot.

This Spud's for You?

Activity

▲ Before the children arrive, prepare a plate for each child with the following items: a few potato sticks, potato chips, a small scoop of potato salad and a little cold potato soup in a paper cup.

▲ Give each child a plate and have them taste each type of potato.

▲ When they have finished tasting, have the children color one of the empty boxes on the graphing grid above the potato product they liked best.

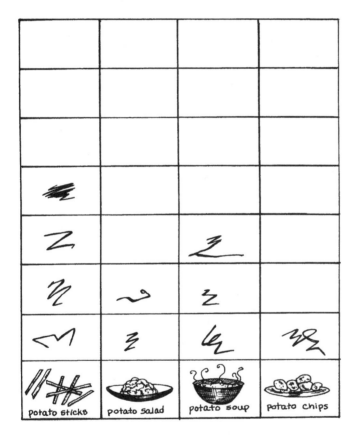

Extension

Graph the meals when children eat the most potatoes: breakfast, lunch, snack or dinner.

More challenging for older children

Make a list of all the food products that are made from potatoes.

Modifications for younger children

Have them sit at a table when eating (it is hard for them to eat with plates in their laps!). Encourage them to taste each product but do not insist on it.

Questions you might ask

▲ What did you like best about each item you tasted?

▲ Who do you know that cooks the foods you tasted?

Interest area
Manipulatives

Science & math principle
Teaches children about the pattern in each group of potatoes.

Science & math skill
Helps children organize what they learned about potatoes.

Science & math attitude
Encourages children to be comfortable with science and math.

▼

You need

different kinds of potatoes (Idaho, Sweet, Russet, New)
large tray
basket

Class-a-fying Potatoes

Activity

▲ Place the potatoes in the basket and the basket on the tray.
▲ Place the tray in the Manipulatives Center.
▲ The children sort the potatoes by size, shape or color.
Note: Save all the potatoes for cooking activities in this section.

Questions you might ask

▲ How are the potatoes the same? different?
▲ Which potato would you like to cook and eat?

Extension

Encourage the children to draw pictures of each group of potatoes.

More challenging for older children

Count the number of "eyes" the potatoes have and sort them from least to greatest, then from greatest to least.

Modifications for younger children

Do this with a small group of children and sort the potatoes by one characteristic, like size. Repeat the activity and sort by another characteristic, like color or shape.

Scaling Potato Heights

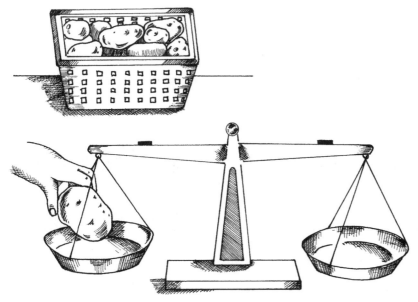

Activity

▲ From the group of seven potatoes, the children select two that will balance the scale when one potato is placed on each side.
▲ If the scale does not balance, have them try again.
▲ Expand the activity by having them select two smaller potatoes.
▲ They place these two potatoes on one side of the scale and see if they can balance them with one larger potato on the other side.

You need

7 potatoes of different sizes
balance scale
basket

Extension

Use weights on one side of the balance scale and potatoes on the other to bring the scale into balance.

More challenging for older children

Use both a balance scale and a numerical scale. Have the children find the number in pounds and ounces (grams and kilograms) for the potatoes that balanced the scales.

Modifications for younger children

Begin with three or four potatoes and add more later.

Questions you might ask

▲ In what other way could you weigh potatoes?
▲ What would happen if you put __ potatoes on one side and __ on the other?

potatoes

Interest area
Science and math

Science & math principle
Teaches children about the growing cycle of potatoes.

Science & math skill
Encourages observation skills.

Science & math attitude
Develops children's curiosity.

You need

1 Idaho potato
1 sweet potato
2 jars (larger in diameter than the potatoes)
toothpicks
water

Spud Rites

Activity

▲ Insert toothpicks around the circumference ("waist") of each potato, just above the middle.
▲ Fill half of each jar with water.
▲ Insert the bottom half of each potato in a jar by resting the toothpicks on the rim of the jar so the potato is partially in the water.
▲ Place both jars in a sunny window. Keep the water level constant.
▲ In about two weeks, the potato will begin to sprout roots and leaves.
▲ As they grow, examine both potatoes and note the similarities and the differences. Add water as needed.
▲ Talk about the parts of the potato plant.

Questions you might ask

▲ How are the potatoes alike? different?
▲ What work do you think the roots do? the leaves?

Extension

Do the activity with other types of potatoes: Rose, Idaho, Cobbler, Early Ohio, Green Mountain, Hebron, Rural and Burbank.

More challenging for older children

Encourage the children to create a picture record of the changes that occur by doing drawings of the different stages of plant growth.

Modifications for younger children

Place the jars out of reach but within sight of the children. Once the potatoes sprout, show the children the potatoes. Talk about the changes each day.

Counting on Potato Sticks

Activity

▲ Put potato sticks in the jar and secure the lid.
▲ The number of potato sticks you use depends upon the level of your class (for example, for three-year-old children, use three or four and for older children, use up to twenty).
▲ Place the jar, the stack of paper, the pencil and the shoe box on the tray.
▲ Put the tray in the Science and Math Center.
▲ The children examine the jar and estimate the number of potato sticks they think are inside. They write down the number on a piece of paper and place it in the shoe box.
▲ After everyone has made an estimate, have each child take one stick from the jar, counting as they go, until all the sticks have been taken and counted. (Use a number line if you think the children are old enough.) Then eat the potato sticks!
▲ Pull the estimates out of the shoe box and compare the estimates to the actual count. Who made the closest estimate?

Questions you might ask

▲ Is there another way we can count the potato sticks in the jar?
▲ How do we know the sticks are made from potatoes?

Extension

Expand the activity by putting some of the potato sticks in a tall olive jar and some in a plastic box. Have the children predict the number of sticks in the jar and in the box. Count them and check their predictions. Have the children compare the number per container (more than, less than).

More challenging for older children

Put more potato sticks in the jar and have them group them by 10s.

Modifications for younger children

Put up to five potato sticks in the jar. Have the children write down their estimations. Count the potato sticks in a group setting (otherwise they will eat them before they count them).

Interest area
Science and math

Science & math principle
Teaches children how to estimate how many are in a group (population).

Science & math skill
Encourages children to compare their predictions and the actual number.

Science & math attitude
Develops children's honesty when comparing results with predictions.

You need

can of potato sticks
clear jar with lid
tray
stack of blank paper
pencil
shoe box

Getting Around Potatoes

Science & math principle

Teaches about the color, size, shape and texture (properties) of potatoes.

Science & math skill

Encourages children to compare their guess to the actual measurement.

Science & math attitude

Develops children's curiosity.

▼

You need

large potato
tray
ball of string
scissors
tape measure

Activity

▲ Place the potato, the string and the scissors on the tray.
▲ Place the tray in the Science and Math Center.
▲ The children look at the potato, estimate how much string they need to go around the potato's length or middle (its "waist") or both, and then cut a length of string.
▲ The children test the string to see how close they were in their estimation.
▲ They can recut the string to the exact length, then measure the string(s) with the tape measure to find the measurement in inches or centimeters.
▲ Be sure to use the word "circumference" when they measure the middle.
Note: Save the potatoes for the cooking activities.

Questions you might ask

▲ What did you learn about the length and width of the potato?
▲ Which was the best tool for measuring?

Extension

Measure other vegetables in the same way and compare the different sizes.

More challenging for older children

Put out a small, medium and large potato for the children to measure and compare string lengths.

Modifications for younger children

Help them cut the string. It will take many attempts before they can cut the string to the size of the potato.

One-Fourth of an Aged Potato

Activity

▲ Cut the potato into fourths, making sure that each quarter has at least one potato eye.
▲ Set potato quarters on paper towels for 1 to 2 days to dry.
▲ Plant one potato quarter in each of the four tires.
▲ Water them when the soil feels dry 3 inches (8 cm) from the surface.
▲ If you want to plant the quarters immediately (without waiting for them to dry), roll them in wood ash from your fireplace. This will keep them from molding.

State the Problem or Question
What will happen if we plant a part of a potato in soil?

State the Hypothesis
Each child predicts what will happen when we plant a quarter of the potato. Write their predictions on a chart.

Method of Research
We will plant one-fourth of a potato in four different tires and record our observations every week. We will water the soil when it is dry. In eighty to ninety days, when the plant that appeared on the soil surface has died, we will dig up the roots and examine them.

Checking the Hypothesis
We will look at our predictions and see if we were correct.

Results
We found that _____ when we planted a fourth of a potato in soil in four separate tires.

Extension

Plant and grow other root plants like beets, carrots, turnips and radishes.

More challenging for older children

Keep a calendar record of the changes in the potato growth.

Modifications for younger children

Let children have their own plants to grow and watch.

Interest area
Science and math

Science & math principle
Teaches children about the growth and development of a potato eye (cycle).

Science & math skill
Enhances observation skills.

Science & math attitude
Helps children learn patience as they wait for the results.

▼

You need

an aged potato (about 2 to 3 weeks old)
Note: You can purchase seed potatoes at a garden store
knife
4 tires filled with soil
trowel
watering can
paper towels
chart tablet and marker

Rubbing Pumpkins

Activity

▲ Place the pumpkin, the container of crayons and the paper on the tray.
▲ The children lay the paper against the pumpkin and rub with the peeled crayon, making a pumpkin-skin rubbing.
▲ Experiment by rubbing different parts of the pumpkin and comparing the results.

Questions you might ask

▲ How are the different parts (surfaces) of the pumpkin different? the same?
▲ Do you think the skin on your body is the same all over or different in some places? Why?

Extension

Make rubbings of other vegetables and compare the results.

More challenging for older children

Use different size pumpkins and compare the results.

Modifications for younger children

Do this activity in a small group. It is hard for most younger children to hold the paper against the pumpkin and to rub at the same time. Have the children select a point on the surface where they want to do their rubbing. Use a push pin or two to hold their paper in place while they rub with the crayon. You can carve the pumpkin later.

Pumpkin Pancakes

mix 4 cups of pancake mix and 1 teaspoon of cinnamon.

Add 2 eggs and 1 cup of pumpkin. Stir well.

Pour small amounts of batter into skillet.

When bubbles form on the pancakes, cook on other side.

Add margarine and honey and eat.

Pumpkin Pancakes Recipe

Interest area
Cooking and snack

Science & math principle
Teaches about irreversible change as the pancakes cook.

Science & math skill
Develops observation skills.

Science & math attitude
Encourages children's curiosity.

▼

You need

Pancakes for Breakfast by Tomie de Paola, or *Pancakes, Pancakes* by Eric Carle
pancake mix, 4 cups (1 L)
2 eggs
ground cinnamon, 1 teaspoon (5 ml)
canned pumpkin, 1 cup (250 ml)
margarine
measuring cups
measuring spoons
stirring spoon
spatula
electric skillet
large bowl
honey or syrup
paper plates, napkins and forks (1 per child)

Activity

▲ Read Tomie DePaola's *Pancakes for Breakfast* or Eric Carle's *Pancakes, Pancakes*.
▲ Combine pancake mix and cinnamon. Add eggs and pumpkin meat. Stir well.
▲ Preheat the skillet to 350º F (180º C). Supervise closely. Add a slice of margarine. After it melts, pour 4-inch (10 cm) circles of batter in the skillet. Cook until bubbles form on the top of the pancakes, then flip them and cook until golden brown.
▲ Serve with margarine, honey or syrup.

Questions you might ask

▲ How many pancakes did we make?
▲ What other things taste like these pancakes?

Extension

Find other pancake recipes, or ask the children to come up with their own recipes to try.

More challenging for older children

Divide the children into two groups. One group makes plain pancakes and a second group makes pumpkin pancakes. Compare the taste. Have them describe how each recipe was the same and each was different.

Modifications for younger children

Use red tape to tape off a "no touch" zone around the hot skillet. Talk about how the hot skillet is dangerous. Put the pancakes aside to cool. Allow the children to do as much of the pouring, mixing and stirring as possible.

Five Round Pumpkins

Activity

▲ Using the felt, make five small orange pumpkins, five yellow pumpkin seeds, one brown loaf of bread, one light-brown stack of pancakes with a yellow pad of butter on top, one orange jack o' lantern and one orange pumpkin pie (use the gray felt to make the pie pan and the light brown felt to make the crust). The children can help make them.

▲ Read and talk about the poem below. Then read it a second time and have the children follow along by putting the felt pieces in place.

FIVE ROUND PUMPKINS

by Sharon MacDonald

Five round pumpkins
In a roadside store
One became a jack o' lantern
Then there were four.

Four round pumpkins
As orange as can be
One became a pumpkin pie
Then there were three.

Three round pumpkins
With nothing fun to do
One was cooked as pumpkin bread
And then there were two.

Two round pumpkins
Basking in the sun
One was cooked as pancakes
Then there was one.

One round pumpkin
One job was left undone
So it was kept to make new seeds
Then there was none.

More challenging for older children

Brainstorm all the things they can do with pumpkin meat and seeds.

Modifications for younger children

Since younger children learn easily through music, demonstrate how to use the felt items while singing the poem to the tune of "Row, Row, Row Your Boat."

Interest area
Group time

Science & math principle
Teaches about the properties of pumpkins.

Science & math skill
Develops communication skills.

Science & math attitude
Encourages children's curiosity.

▼

You need

orange, yellow, brown, gray and light brown felt
scissors
markers or pen

Questions you might ask

▲ What happened to each pumpkin?
▲ Why was it important to keep one for new seeds?

Extension

Encourage the children to write their own pumpkin poems.

Planting Pumpkin Seeds

Interest area
Group time

Science & math principle
Teaches about the plant life cycle.

Science & math skill
Develops observation skills.

Science & math attitude
Helps children learn to be patient.

▼

You need

2 pumpkin seeds for each child
soil in a container
containers for planting the seeds (empty milk cartons work well)
spray bottle filled with water
scoops or spoons
tray

Activity

▲ Place the soil, spoon, two seeds, the container and spray bottle on a tray.
▲ Read the following poem below and hold up the package of pumpkin seeds that the children will be planting.

THE PUMPKIN SEED
by Sharon MacDonald

The package said pumpkin seed.
I know it's true 'cause I can read.

I put the seed into the ground,
And watered it carefully all around.

Soon, the seed sprouted and started to grow.
But how it knew to—I'll never know.

The vine grew outward very low.
"Up" wasn't the way it wanted to go.

Flowers blossomed—some here, some there.
And small orange balls were everywhere.

As fall slowly ceased to be,
The orange balls smiled at me.

I understand that pumpkin smile.
'Cause each one of them knew all the while,

That in each seed hidden secretly,
Is the pumpkin knowledge of what to be.

▲ After you have talked about the poem, invite the children to plant two pumpkin seeds, following the directions.
▲ The children track the growth of their pumpkin plants on a calendar.
▲ Encourage them to draw pictures of the growth and change.
▲ After the seeds have sprouted, send them home with the children to be planted in their yards. For pumpkins to survive, they can live only a limited time in the milk carton (no more than a week or so after sprouting). They must be transplanted to soil if they are to grow. Do this activity in the spring so the pumpkins can grow throughout the summer.

you need: Planting a Seed

Dirt

containers

seeds

water

put the dirt in the container

put the seed in the dirt.

Water the seed.

Questions you might ask

▲ Why do you think it is a good idea to plant two seeds?
▲ What do you think will happen to your seeds?

Extension

Transplant the pumpkin plants outside in a school garden. You'll have a pumpkin patch for the next year's children.

More challenging for older children

Keep track of the plant by placing a ruler inside the milk carton to measure growth.

Modifications for younger children

The spray bottle may be difficult for them to use. Substitute a misting bottle that operates with a push-down rather than hand-squeeze action.

Carving "Lighty" Pumpkins

Interest area
Group time

Science & math principle
Teaches about cause and effect as they carve and help light the pumpkins.

Science & math skill
Develops observation skills.

Science & math attitude
Encourages the children to cooperate as they decide on a design.

▼

You need

large pumpkin
candle, 4" (10 cm) in diameter and 4" (10 cm) tall
carving knife
large bowl or pan
newspaper
heavy-duty spoons
paper and pencils or crayons

Activity

▲ Read the story below.
▲ Divide the children into groups of three or four. Each group draws a design of how they think a pumpkin should be carved to give off lots of light.
▲ Place the pumpkin on the newspaper and cut off the top. (Adults do most of this step.) The children clean out the pumpkin.
▲ Place each group's designs on the wall and have all the children vote on the design they think is best. When the design is chosen, draw it on the pumpkin and carve it. (Adult does the carving.)
▲ Put the candle inside the pumpkin and light it. Turn off the classroom lights to test the children's design.

JACK'S LANTERN

by Sharon MacDonald

Once there was a man named Jack. He lived on a farm. It was very dark on Jack's farm at night. Every night, Jack had to go out to the barn to feed his cow. It was a long walk to the barn and Jack did not like to walk in the dark. It scared him. He wished he had some kind of light.

He tried using a candle, but when he carried the candle as he walked, the wind blew it out. So he tried to cover the candle with a piece of paper, but the flame burned up the paper. He almost burnt his fingers! Jack thought about his problem.

Jack grew pumpkins on his farm, too. One day, when Jack was picking pumpkins to sell at the market, he found one with a long, narrow slit up the side. He knew he could not sell that pumpkin, so he took it inside the house. He cut off the top and he scraped out the seeds and pumpkin meat. He set aside the seeds in a bowl to wash and cook later.

Jack looked inside the pumpkin to make sure he had scraped out all the meat and the seeds when he noticed he could see light coming through the narrow hole. Jack got an idea. If he put a candle inside the pumpkin, it would shine out through the slit. He would have candle light to see when he took his long walk to feed his cow each night. But Jack thought by cutting more slits in the side of the pumpkin, he would have even more light. So he cut more slits in the side of his pumpkin. He put in the candle, and he lit it. He put on the top. The candle light shined through the slits. Jack was happy and excited about trying out his candle in the pumpkin when it got dark.

That night he lit the candle inside the pumpkin and he put on the top. He carried the pumpkin outside. The wind did not blow out his candle. Now, Jack had light that would let him see in the dark as he took his long walk to the barn to feed his cow. He was not scared any more, either. Jack had made his very own lantern!

Jack made lanterns for his family and friends so they could see when they took walks to feed their cows. When other people wanted pumpkin lanterns, they would go to Jack. They called him "Jack-of-Lanterns" because he made the special lanterns that gave off the light.

As years went by, people never forgot how they learned to carve pumpkins and make pumpkin lanterns. They called their carved pumpkins Jack O' Lanterns in memory of Jack. Maybe that is why they call them that today, especially on Halloween!

Questions you might ask

▲ Why does the pumpkin lantern help us see in the dark?
▲ What other ways could we carve the pumpkin to get light?

Extension

Use the seeds to roast and to plant. Use the meat to make a pumpkin pie with the class.

More challenging for older children

Encourage the children to make up a different story about how the name Jack O' Lantern was started.

Modifications for younger children

Carve the pumpkin in small groups so each child can help design and clean the pumpkin.

Interest area
Manipulatives

Science & math principle

Teaches about patterns as children examine the relationship of one pumpkin to another.

Science & math skill

Encourages children to organize their information about pumpkins.

Science & math attitude

Encourages children to seek more information (curiosity).

▼

You need

6 different-size pumpkins

Questions you might ask

▲ How is the biggest like the smallest?
▲ Can you think of another way to put the pumpkins in order?

Pumpkins in a Row

Activity

▲ The children put the pumpkins in order by size from the smallest to the largest, and then from the largest to the smallest.

Extension

Seriate by the weight of the pumpkins. Note: this is a very important developmental task. It is also very hard for some children, as well as adults, to do. One must pay exclusive attention to an "embedded" quality (like weight, or some other characteristic) while ignoring sensory data (like visual information, for example) which may suggest a different order to the seriation.

More challenging for older children

Seriate the pumpkins by the degree of color: from the lightest orange color to the darkest orange color.

Modifications for younger children

Start with three pumpkins of very different size; add more as the children develop their skills. Use a silhouette board first, so the children just need to match the pumpkin to the silhouette.

The Pumpkin Sort

Activity

▲ The children sort the pumpkins by those that have stems and those that do not.
▲ Then they sort them by those that have deep ridges and those that have shallow ridges.
▲ Then they sort them by those that are darker in color and those that are lighter.
▲ Ask the children to talk about other ways they could sort them.

Interest area
Manipulatives

Science & math principle
Teaches about patterns as children examine the relationship of one pumpkin to another.

Science & math skill
Teaches children to organize their information by sorting.

Science & math attitude
Encourages a desire for more knowledge (curiosity).

▼

You need
5 or 6 pumpkins

Extension

Ask the children to write descriptions of three pumpkins and then find what is common to the three descriptions.

Questions you might ask

▲ Why don't the pumpkins all look exactly the same?
▲ Is there something that is exactly alike about all the pumpkins?

More challenging for older children

Encourage the children to sort the pumpkins by those that are dark, have stems and are deep ridged; then by those that are light, lack stems and are shallow ridged. Ask if there are any that haven't been sorted. Describe them.

Modifications for younger children

Sort only one way the first day. Each following day, sort a different way.

Watching a Pumpkin Decay

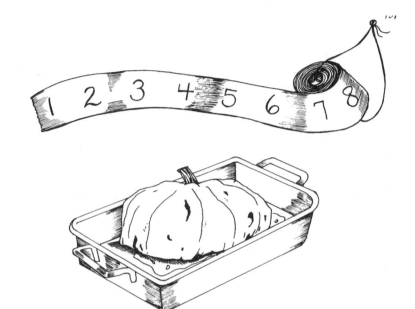

Interest area
Science and math

Science & math principle
Teaches about change as children watch the pumpkin rot.

Science & math skill
Encourages children to compare what is happening to what has already happened.

Science & math attitude
Helps children develop patience.

You need

dishpan
medium-size pumpkin
adding-machine tape
string
scissors
push pin or tape
chart tablet and marker

Activity

▲ Before the children arrive, tie a string through the spool of the adding-machine roll and tie it off. Hang from a push pin in the wall.

▲ Place the whole (uncut) pumpkin in the dishpan.

▲ Ask each child to predict how many days it will take for the pumpkin to decay (rot).

▲ Ask them to think about how the pumpkin will look when it has completely decayed (it will collapse under its own weight because of a loss of water). Write down the children's predictions and post them on the wall beside the roll of adding-machine tape.

▲ Each day, check the pumpkin and write a number on the tape to keep track of the number of days.

▲ When the pumpkin has decayed and is dry enough to pick up, check the children's predictions.

Note: Plan plenty of space for the adding-machine tape. It could take as long as six months for the pumpkin to rot and dry. It will seem like nothing is happening for a very long time. When you see moisture oozing from the bottom of the pumpkin, move it outside to a protected area. It will begin to smell. The smell lasts for about a month and then the pumpkin starts to dry. To speed up the drying time, use a baster and draw off the liquid residue that accumulates in the bottom of the dishpan. If you try to pour off the liquid, the pumpkin will loose its shape and prematurely collapse. Try not to move it so it retains its pumpkin shape as long as possible.

Questions you might ask

▲ Why did it take so long to dry?
▲ Where do you think the liquid in the bottom of the dishpan came from?

Extensions

Keep a weekly class journal of the changes that are observed and then, as the pumpkin begins to change visibly from one day to the next, keep the journal on a daily basis. Take photographs of the decay process at significant intervals, when the changes are most noticeable.

More challenging for older children

Compare the decay of a whole pumpkin to that of a pumpkin that has been cleaned out or opened in some way.

Modifications for younger children

Ask each child to poke the pumpkin with one finger to test the amount of decay (feel how soft the skin is).

Measuring Pumpkin Parts & Places

Interest area
Science and math

Science & math principle
Teaches about the properties of pumpkins.

Science & math skill
Teaches children to compare the circumferences of pumpkins.

Science & math attitude
Encourages children to be honest as they check their predictions.

▼

You need

medium-size pumpkin
string
scissors
black marker
large tray
four blank cards

Activity

▲ Before the children arrive, use the black marker to make circumscribing lines around the pumpkin at four points: the "equator" (horizontally); the top third; the bottom third; and the "poles" (vertically). These are reference lines. The children will use them as measuring lines when the time comes for them to test their predictions (the children will place their string on top of these lines).

▲ Make the four cards, each one showing the pumpkin with only one of the four lines mentioned above (see illustration).

▲ Cut four lengths of string to fit around the pumpkin at the points described above.

▲ Place the four lengths of string, the picture cards and the pumpkin on the tray.

▲ Put the tray in the Science and Math Center.

▲ The children pick up a piece of string and guess where it might fit. Then they match their string selection to one of the four cards on the tray.

▲ After the children have predicted all four string positions and matched the string with the appropriate card, have them take each string and test it on the pumpkin, measuring over the black reference lines. For example, a child picks up a string and decides it will go around the middle of the pumpkin. She will then put the string on the card that depicts the pumpkin with the string around the middle. After she has placed all four strings on cards, she will take each string and see if it fits around the pumpkin at the points she predicted.

Questions you might ask

▲ Why are the strings different in length?
▲ What else can you use to measure the pumpkin?

Extension

After they finish the activity have them measure the strings with a tape measure and record their findings on paper.

More challenging for older children

Encourage children to illustrate their measured pumpkin.

Modifications for younger children

Measure the pumpkin at the equator only. Use a small pumpkin so that the children can turn it easily to measure with the lengths of string.

Weighty Pumpkins

Interest area
Science and math

Science & math principle
Teaches children about the diversity of sizes of pumpkins.

Science & math skill
Teaches children to compare weights.

Science & math attitude
Encourages children to keep an open mind as they guess and then check their guesses.

▼

You need

kitchen scale (that measures in pounds and ounces or grams and kilograms)
4 or 5 different-size pumpkins
dry-erase marker
dry-erase eraser

Activity

▲ The children take turns picking up each pumpkin to feel and guess its weight.
▲ Using a dry-erase marker, they write (or you write for them) the weight they guessed on the pumpkin.
▲ Then they place the pumpkin on the kitchen scale to see how close their guess was to the actual weight.
▲ They erase the guess and move to the next pumpkin, repeating the procedure.
▲ This can be a group activity where you select children or ask for volunteers, or you can place this activity in the Science and Math Center for independent use.

Questions you might ask

▲ How much does each of the pumpkins weigh?
▲ What makes them heavy?

Extension

Use a bathroom scale to compare the results of weighing all the pumpkins. Discuss the accuracy of the results.

More challenging for older children

Encourage the children to guess without picking up the pumpkins. Making predictions about weight without giving them a chance to lift the pumpkin makes the task of accuracy much more difficult.

Modifications for younger children

Use two or three pumpkins and have them say their guess outloud. Help them weigh the pumpkin to see if their guess was right.

How Many Seeds Does a Pumpkin Hold?

Activity

▲ Ask the children to predict how many seeds they think are inside the pumpkin.
▲ Mark down their predictions.
▲ Check their predictions by cleaning out the pumpkin and counting the seeds.

State the Question
How many seeds are inside the pumpkin?

State the Hypothesis
Each child will predict how many seeds are inside the pumpkin. The number will be recorded beside the child's name on a large sheet of paper and posted in the classroom.

Method of research
The children will clean out the pumpkin, wash the seeds and then count the seeds.

Checking the Hypothesis
We will look at our list of predictions and see if we were correct.

Results
We found _____ seeds inside our pumpkin.
Note: This is a great opportunity to count by grouping by tens.

Questions you might ask

▲ Do you think a small pumpkin would have the same number of seeds?
▲ Why do you think a pumpkin has so many seeds?

Extension

Use other vegetables with accessible, large seeds.

More challenging for older children

Divide the class into small groups of three or four and give each group a pumpkin. Have them make their own predictions and check their results. Compare the results in each group.

Modifications for younger children

Use the hybrid pumpkin (it is about the size of an orange). It has only a few seeds and younger children will not lose interest as they count.

Interest area
Science and math

Science & math principle
Teaches about population as children examine the number of seeds in the pumpkin.

Science & math skill
Teaches children to compare their estimates to the actual number.

Science & math attitude
Encourages children's curiosity.

You need

large pumpkin
knife
large bowl
newspaper
large sheet of paper with each child's name on it
chart tablet and marker

Interest area
Art

Science & math principle

Teaches about interaction between rocks, paper and paint.

Science & math skill

Develops observation skills.

Science & math attitude

Encourages children's curiosity.

▼

You need

mailing tube 2'–3' (60–90 cm) long and 2"–3" (5–8 cm) in diameter (Tape bottom or purchase kind that does not have a removable bottom.)

paper cut to fit inside the cylinder (paper should be as wide as the length of the mailing tube and about 9"–10" or 22–25 cm long)

plastic plate or tray with 3 divisions (save frozen dinner trays)

tempera paint, 3 colors

3 rocks, about 1" (2.5 cm) around

3 plastic spoons

smock

newspaper

shoe box without a lid

Rock Rolling

Activity

▲ Children choosing this activity put on the smock, cover the workspace with newspaper, place the three-compartment paint tray on the table, pour in the three colors of tempera paint (one color in each compartment) and put one rock and one plastic spoon in each compartment.

▲ The children put the paper into the mailing tube. Using the spoon, they put one, two or all three paint-covered rocks in the tube, then put on the lid (you might need to secure the lid further with tape).

▲ The children shake or roll the tube.

▲ When they are through shaking and rolling, they remove the lid and dump the rocks into the shoe box (the box is simply a place to accumulate the rocks rather than having them on the floor or the table).

▲ Pull out the paper from inside the tube and examine the rock designs in tempera paint. A child doing this activity can repeat the exercise, or clean up and put his paper aside to dry.

Questions you might ask

▲ How else can you make a similar design?

▲ How would it feel to be a rock inside the tube, sliding and rolling and bouncing around?

Extension

Do this activity using a small can, such as a tennis ball or potato chips can.

More challenging for older children

Ask the children to create a story to go along with their rolling-rock design.

Modifications for younger children

Have two children do this together at the same time. They can roll the tube back and forth to each other on the floor, rather than trying to hold and shake it.

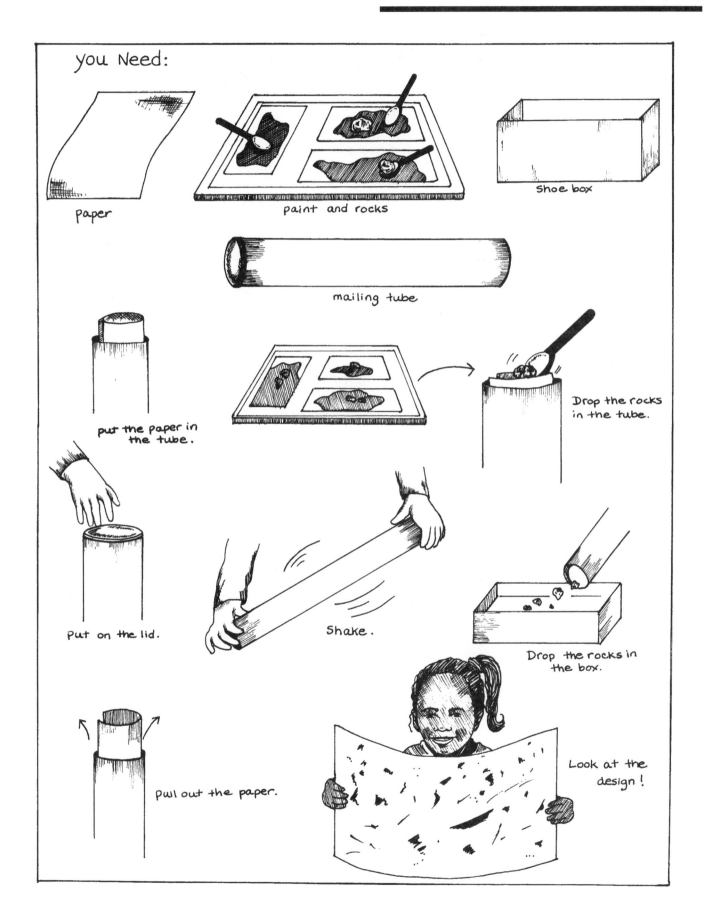

you Need:

Paper

paint and rocks

Shoe box

mailing tube

put the paper in the tube.

Drop the rocks in the tube.

Put on the lid.

Shake.

Drop the rocks in the box.

Pull out the paper.

Look at the design!

Rock Collage

Interest area
Art

Science & math principle

Teaches about cause and effect as children glue the rocks on the cardboard.

Science & math skill

Develops communication skills.

Science & math attitude

Develops respect for materials and tools used.

You need

cardboard pizza circles (most pizza shops will donate them)
glue
rocks of various sizes, shapes and colors
hair dryer

Activity

▲ Place the hair dryer close to the work area.
▲ The children create a design by arranging rocks on the cardboard pizza circles. They can also build rock layers.
▲ They use the glue to attach the rocks to the cardboard pizza circles or to each other.
▲ They use the hair dryer (with supervision) to speed the glue drying when they have finished (speed drying is especially helpful with layered rock collages, keeping all of the rocks in place).
Note: If the children dry as they build, they can create much taller structures.

Questions you might ask

▲ Can you tell someone how to design a collage like your design?
▲ Did the hair dryer make your work easier? How?

Extension

Display several rock mosaics for the children to see. Discuss how the artist created a picture mosaic by covering the entire surface. Give the children colored aquarium rocks to create their own mosaics. They work individually or in groups.

More challenging for older children

Ask all the children to work together on a group rock collage.

Modifications for younger children

Do not use the hair dryer. Younger children seldom wish to layer their rocks and, due to the pace at which they usually attach and arrange their rocks, speed-drying is not important.

Rock Story

Activity

▲ Draw three large circles on each piece of paper.
▲ Read or tell the children the *Rock Day* story below. Discuss what went on in the story, focusing on parts of the story the children really liked.
▲ Then divide the children into groups of three. Give each group nine rocks and a piece of paper with the three circles drawn on it.
▲ The children work in teams to divide the rocks by color and texture, and then by size, just like the children did in the story.

ROCK DAY

by Sharon MacDonald

Meet Meg and Sasha. They are best friends and they live next door to each other. One afternoon when Meg and Sasha were playing in Sasha's backyard, they found a place near the fence with all kinds of rocks scattered around. They started to play with the rocks, popping them out of the ground with their fingers.

Meg and Sasha had a friend named Robert. He called himself a "rock hound." Sasha and Meg had no idea what "rock hound" meant, but it sounded like a fun thing to be. They had seen Robert's rock collection, and they wanted collections of their own.

They dug around in the dirt, popping up the rocks they liked. They decided to find black rocks for Robert, since they were his favorite kind. But they wanted rocks that were just the right size, just the right shape and most of all, just the right color. Sasha decided she liked the ones with red in them the most. Meg liked mostly the white ones. They found rocks with black on them for Robert.

Meg looked at her rocks and felt them. She did not like the bumps on her rocks. Meg liked smooth rocks. Sasha liked her red rocks, but one felt sandy when she rubbed it and she did not like that. She liked her rocks bumpy and lumpy. They both knew that Robert liked black rocks, but black and sandy rocks would be best so he could draw on the sidewalk with them. They were not happy with the way they had divided the rocks. What to do?

They decided to divide them by size instead. Robert collected big rocks, so they decided to give him the biggest ones. Meg wanted medium-sized rocks; Sasha liked the small ones. Now they had it right. This was the way to separate the rocks. They each had three. They wanted to take them to Robert's house, but only Meg had a pocket. What to do?

Meg wrapped her three rocks in a piece of paper she found in the yard. Sasha found a big leaf and rolled up her rocks inside it. They could not find anything to put Robert's rocks in so they dropped them in the bottom of Meg's pocket. Meg also put her own rocks, wrapped in paper, and Sasha's rocks, wrapped in the leaf, into the same pocket. They got on their bikes and rode down the street to

Interest area
Group time

Science & math principle
Teaches children to recognize patterns in the rocks.

Science & math skill
Teaches children to organize information by sorting rocks.

Science & math attitude
Encourages children to cooperate.

You need

9 small rocks for each group of 3 children
1 piece of paper for each group

Robert's house. They couldn't wait to share their rocks with Robert!

When they saw Robert, Meg pulled the leaf from her pocket, then the paper, but all the rocks fell out and got mixed up at the bottom of her pocket. What to do?

Robert, Sasha and Meg sat and thought. "Let's put all the rocks in a pile," said Sasha. "We can each take one until they're all gone." So each child chose a rock until all the rocks were gone. Everyone was happy with their rocks. Robert told Meg and Sasha that they were very good rock hounds. Now they knew what rock hound meant! They liked being rock hounds, just like Robert.

Questions you might ask

▲ What size, color or shape of rock do you like?
▲ Why did all the rocks got mixed up in Meg's pocket?

Extension

Ask the children to draw pictures for the story and place them in a big book format.

More challenging for older children

Ask the children to work in small groups to describe the nine rocks in the story.

Modifications for younger children

Work with the small groups separately so you can help them sort the rocks.

Rock Grouping

Activity

▲ Make grouping circles by cutting two lengths of yarn 24 inches (60 cm) long. Tie the ends of each piece together, creating two circles.

▲ Place the rocks in the container and the yarn loops on the tray.

▲ The children make two yarn circles and sort the rocks into two groups by color, size, shape or texture. For example, they could sort by color and size; color and shape; color and texture; or size and shape.

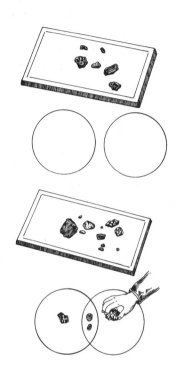

▲ After they understand the grouping process, overlap the circles to create three grouping areas. The center area is for rocks that have both grouping characteristics in common. For example, the children would separate brown rocks from black rocks in the two grouping circles; then in the middle, or third group, they would group rocks that are both black and brown (this is a foundation skill for Venn diagrams).

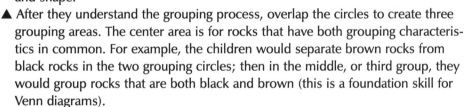

Extension

Gather rocks that are similar in shape, size, color and texture thereby making it more difficult to discriminate between characteristics.

More challenging for older children

Divide the class into small groups. Give each group a small handful of rocks and yarn circles so they can sort the rocks by color, size, texture or shape. Then they can compare how they sorted their rocks to how other groups sorted their rocks.

Modifications for younger children

Draw circles on a large sheet of paper or cut tape to make grouping circles on a tray. Demonstrate the activity in small groups (before putting it in the Manipulatives Center).

Interest area
Manipulatives

Science & math principle

Teaches about the many kinds of rocks (diversity).

Science & math skill

Teaches children to organize their knowledge by grouping rocks.

Science & math attitude

Encourages children's curiosity.

You need

8 to 10 rocks of different size, shape and color
small container
thick colored yarn, such as pony-tail yarn, or light-weight rope
large tray

Questions you might ask

▲ What other ways can you group the rocks?

▲ Which kind of rocks do you like the best? smooth, rough, black, gray, big, little, round or flat?

Rock Sorting

Interest area
Manipulatives

Science & math principle
Teaches about the relative sizes of rocks (scale).

Science & math skill
Teaches children to organize their knowledge by seriating the rocks.

Science & math attitude
Encourages a desire for more knowledge.

▼

You need

3 shoe boxes
marker
2 large, 4 medium and 6 small rocks in a container
scissors
tray
index cards
pencils

Questions you might ask

▲ Would the big rock fit in the small hole? Why not?
▲ Would the little rock fit in the big hole? Why?

Extension

Go on a rock hunt to find large, medium and small rocks to test in the sorting boxes.

Activity

▲ Make sorting boxes by cutting a large hole in the first box lid (large enough for the largest rock to pass through), a medium-size hole in the lid of the second box and a small hole in the lid of the third box. Write large, middle-size (or medium) and small beneath each corresponding hole on the box lids, and replace the lids on the boxes.
▲ Place the three boxes on a tray with the rocks in the container. Put the index cards and pencils beside the tray.
▲ The children choose a rock and decide into which hole the rock will fit. They test their guesses by trying to push the rocks through the holes.
▲ When all of the rocks have been sorted, the lids are removed and the number of rocks in each box are counted.
▲ Each child doing the activity documents her research by writing the numbers of rocks in each group on her index card. Children who cannot write numbers yet can make marks to represent their findings.

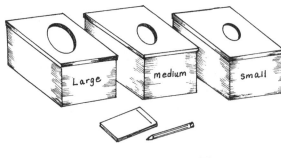

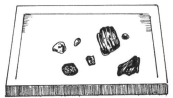

Where does the rock belong?

record your findings

More challenging for older children

Create more sorting categories (large, medium, small, very small and miniature) and more boxes into which the rocks will be sorted.

Modifications for younger children

Start off this activity with two boxes. Transition to three boxes as the children become familiar with the sorting process.

Rock 'n Roll Band

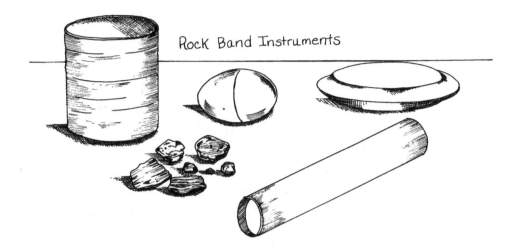

Rock Band Instruments

Interest area
Music and movement

Science & math
principle
Teaches children to listen for patterns of sounds.

Science & math
skill
Develops observation skills.

Science & math
attitude
Encourages children's curiosity.

▼

Activity

▲ Encourage the children to create a rock 'n roll band by making and using the following instruments:

(1) a rock drum by placing five or six small rocks of different sizes in a coffee can, then sealing the top with tape or a hot glue gun. The children tap on the lid, making the rocks bounce in the bottom.

(2) a rock tambourine by placing six or seven small rocks inside two paper plates, with the tops opposing (see illustration); staple or glue together.

(3) a rock shaker by placing ten miniature rocks inside the small container and securing the lid.

(4) a rock 'n roller by placing four small rocks inside the mailing cylinder, then sealing it with glue or tape.

(5) rock "cymbals" by hitting two of the larger rocks together.

▲ Put on a music tape and encourage the children to follow the rhythm. All the instruments are shaken except the rock cymbals.

Extension

Compare the rock instruments to the musical instruments used by real bands.

More challenging for older children

Give the children a variety of rocks and a wide variety of different containers. Let them create their own band sounds.

Modifications for younger children

Make sure the containers are well-sealed and that the children understand how to use the musical instruments. Be a part of the band for a few sessions to model the use of the instruments.

You need

variety of rocks, some 5"–6" (12–15 cm), some less than 3" (8 cm) and many miniatures
empty coffee can with a plastic lid
small plastic container with secure lid
2 sturdy paper plates, such as Chinet
3' (1 m) mailing tube, 2"–3" (5–8 cm) in diameter
tape
hot glue gun (for use by adult only)
stapler

Questions you might ask

▲ How are the sounds of the instruments the same? different?

▲ What did we use in all the instruments that is the same?

Rock Washing and Polishing

Interest area
Sand and water

Science & math principle
Teaches about interaction between water, sandpaper and rocks.

Science & math skill
Develops observation skills.

Science & math attitude
Encourages children to be curious.

You need
small dishpan or water table
water
variety of large rocks
toothbrush
sandpaper of different
    coarseness (rough,
    medium, fine, extra fine)
2 trays
1 large and 1 small towel

Activity
▲ Place the rocks on a small tray.
▲ Place the large towel on the workspace.
▲ Fill the dishpan or water table half full of water. If using a dishpan, put the dishpan on the large towel.
▲ Put the toothbrush in the water.
▲ Place the small towel beside the dishpan.
▲ Place the sandpaper on the other tray and set it beside the work area.
▲ The children choose a rock, submerge it in the water, brush it clean with the toothbrush and dry it with the small towel.
▲ They choose a piece of sandpaper and polish the rock. They experiment with the effects of polishing their washed rock with sandpaper of different coarseness.

Questions you might ask
▲ What did the sandpaper do to your rock?
▲ Why did you choose that particular rock? What did you like about it?

Extension
As they use the sandpaper on the rocks, collect the accumulated grit and sand in a box. Examine it with a magnifying glass.

More challenging for older children
Compare the rocks the children have cleaned. Decide how they are alike and how they are different. Why?

Modifications for younger children
Use a small pan with enough water to cover the bottom. The rocks will not sand well if they are wet, so encourage the children to dry them well.

Rock Filtering

Activity

▲ Make a home-made sieve by puncturing holes in an aluminum pie tin or butter tub (to puncture the butter tub, heat the tip of an ice pick on the stove first). This step is done by adults only.

▲ Mix very small rocks and sand together. Place the sieves, scoops and this rock-and-sand mixture in the sand and water table or in the Science and Math Center in a large dishpan.

▲ The children use scoops to pour sand into the sieves and to examine the rocks they find after the sand has gone through.

▲ Encourage them to notice which sieves caught the most rocks and which let many of the rocks pass through.

▲ Have the children tier or seriate the sieves by hole sizes to see what they can catch at each level and what comes out of the last sieve at the bottom.

Extension

Try this activity with moist sand. Give the children spatulas and spoons. Have them mash the sand through the sieve.

More challenging for older children

Let the children design and make their own sieves. Give them a wide variety of materials to work from. Supervise well.

Modifications for younger children

Make sure the handles on the sieves and scoops are manageable for younger children. If necessary, wrap them with tape to make them easier to hold and less slippery.

Interest area
Sand and water

Science & math principle
Teaches about interaction of rocks, sieves and sand.

Science & math skill
Teaches children to compare the results of working with different sieves.

Science & math attitude
Encourages children to be more comfortable with science and math.

You need

large pasta sieve
tea strainer
colander
homemade sieve (see instructions)
sand
very small rocks, different sizes and colors
large tray or sand table
scoops
dishpan

Questions you might ask

▲ What is similar about the sieves? what is different?
▲ How many rocks do you think are buried in all the sand?

Breaking a Rock-in-a-Sock

Interest area
Science and math

Science & math principle
Teaches cause and effect when children see what happens to the rocks.

Science & math skill
Develops observation skills.

Science & math attitude
Teaches respect for tools used in science and math.

▼

You need

several large rocks in a basket
large supply of old tube socks in a basket
hammer
goggles
chopping board
towel
tray
magnifying glass

Activity

▲ Fold the towel and place it on a table.
▲ Put the chopping board on top of it (the towel absorbs the hammering noise).
▲ Put the hammer, goggles, rocks and socks within easy reach of the children.
▲ The child takes a rock from the basket, drops it in a tube sock, puts on the goggles, puts the sock with the rock on the chopping board and hits the rock in the sock with the hammer.
▲ When the rock breaks, the child shakes out the contents from the tube sock and places the rock pieces on the tray for later examination under a magnifying glass.

Note: This activity must be closely supervised. Be sure to examine the tubesocks frequently and replace the ones with holes. Hammering rocks in socks tears the fabric, making holes in short order. Pieces can easily fly from the holes and cause injury.

Questions you might ask

▲ What could you use to put the rock back together?
▲ What other tools might you use to break the rock?

Extension

Use the broken rock pieces for sorting and matching games, for collage material and for starting a classroom rock collection.

More challenging for older children

Give them different rocks of different hardness. See if they can discover what hard-to-break rocks and easy-to-break rocks look like. How are hard and soft rocks alike? How are they different?

Modifications for younger children

Select limestone rocks that are easier for younger children to break. Get a hammer without a claw for them to use.

Rock Graphing

Activity

▲ Make a graphing grid on the window shade by laying out four-inch squares: eight horizontally across the top with twelve rows from the top of the shade to the bottom. Use a permanent marker and a yardstick to lay out straight lines. When finished, you should have 96 squares (see illustration).

Note: Making a graphing grid will take time and patience. In a science- and math-oriented center, however, the grid will be used over and over again.

▲ Take the children on a rock hunt. Each child carries a shoe box (paper or plastic sacks tear too easily) to put collected rocks in.

▲ When you return from the hunt, roll out the window-shade graphing grid. Gather the children and ask them to select two of their rocks and put the rest in the box behind them. They will place their two rocks on the graphing grid when their turn comes.

▲ Start the activity with a child placing a rock in any one of the boxes along the bottom row of the grid by the characteristic that the child finds to be the most prominent (for example, "lumpy," or "brown"). Then ask if anyone has a rock with that same characteristic. That child places her rock above (not beside) the first rock. Other children select a rock that matches the characteristic mentioned by the first child and place it on the graphing grid above the other two. Not all of the children will have such a rock.

▲ Ask a second child to select a rock and mention the predominant characteristic. Have her place it on the grid next to the first column, and to continue the matching until all of the children who have a rock with that characteristic have put it on the grid.

▲ After all the children have placed their two rocks on the grid, talk about the rock groupings.

Note: One objective of this activity is to have rocks grouped in vertical columns by some of the children, but not necessarily all of them, based on a similar characteristic. Some children may not think they have a rock with a similar characteristic to others in the group. That being the case, they will not participate on the first round. That's okay. A second objective is to have the children notice similarities and differences, but it is not necessary that they respond to the activity. For example, a child might respond to a friend but not to another child, even though there is little similarity between their friend's rock and their own. This dissimilarity should not be pointed out by the teacher, since the real purpose of this kind of activity is not to find a "correct" answer but rather to learn about diversity. The third objective of this activity is problem solving. For example, if you find that you do not have enough empty boxes on the grid for all of the rocks being grouped, let the children brainstorm solutions. Such an approach encourages improvisation and creativity, prized qualities in all of us!

Interest area
Science and math

Science & math principle
Teaches about the many kinds of rocks (diversity).

Science & math skill
Teaches children to communicate by graphing their information.

Science & math attitude
Teaches respect for nature.

You need

rocks
shoe box, 1 per child
old roll-up window shade, 3" (8 cm) wide by 3'–6' (1–2 m) long
permanent marker
yardstick

rocks

Questions you might ask

▲ What do you see that is alike about all of these rocks?

▲ What would the earth look like if all rocks were lumpy and white?

Extension

Place the graph in the Science Center and encourage the children to take their shoe box rock collection and graph their rocks.

More challenging for older children

Give the children graph paper you have drawn with large boxes. Encourage them to place their rocks on the graph paper.

Modifications for younger children

Gather the children in groups of three or four. Have each child put one rock she collected in a basket. Help the children notice similarities and place the rocks on the graph. Let them explain why the rocks are alike.

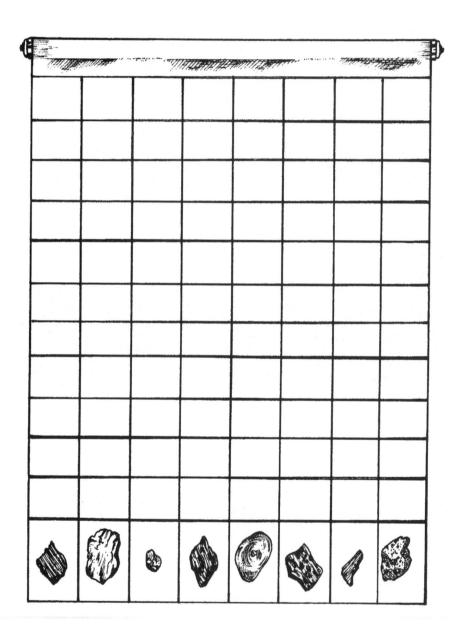

Rock Weighing

Activity

▲ The children predict how many smaller rocks it will take to make the balance scale level when there is a large rock on one side of the balance scale.

State the Problem
How many smaller rocks will it take to level the balance scale when the large rock is placed on one side of the scale?

State the Hypothesis
We believe it will take _____ small rocks. (You can use individual predictions or one by the whole class.)

Method of Research
We will place a large rock on one side of the balance scale, and we will add rocks to the other side of the scale until the scale is balanced. We will add up the number of smaller rocks it took to balance the scale.

Checking the Hypothesis
We will compare our prediction(s) with the actual results.

Results
It took _____ rocks to level the balance scale.

Questions you might ask

▲ What would it feel like to be in the tub with the smaller rocks?
▲ If we had an even larger rock on one side, how many smaller rocks do you think it would take to balance the scale?

Extension

Move the activity to a numeric scale (often used in the kitchen for food weighing). Do the same activity making the same weight predictions and comparisons in numbers.

More challenging for older children

Let them set up their own experiment and collect their own rocks. Have them compare their findings with the class experiment and with the other older children's rock experiments.

Modifications for younger children

Do this first by placing the large rock in a child's hand and a few smaller rocks in the other hand. Let them grasp the idea of balancing things. Once they understand weight and balance, let them move to the balance scale.

Interest area
Science and math

Science & math principle
Teaches about scale as children measure different amounts.

Science & math skill
Teaches children to compare results to estimates.

Science & math attitude
Teaches respect for the tools of science and math.

You need

1 large rock (about the size of a lemon)
12 to 15 small rocks
balance scale

Tap Dancing Shoe Printing

Interest area
Art

Science & math principle
Teaches children about cause and effect as they print.

Science & math skill
Encourages children to develop organization skills.

Science & math attitude
Develops children's curiosity and their desire for knowledge.

▼

You need

1 old shoe
washer (to be used as a tap)
hot glue gun
newspaper
paper
stamp pad
tray

Questions you might ask

▲ What would happen if you tap the shoe on the paper to music?
▲ How many taps can you make before you run out of ink?

Activity

▲ Make a tap shoe by hot gluing (adult only) a washer to the toe of the shoe.
▲ Set up the area by covering the table with newspaper.
▲ Place the tap shoe, paper and stamp pad on a tray.
▲ The children press the toe of the tap shoe into the stamp pad and then "tap" the paper to create a design.

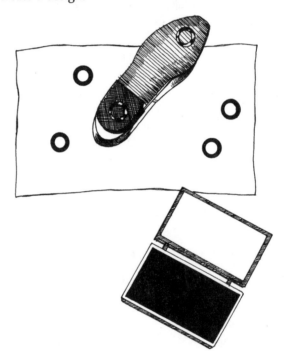

Extension

Hot glue different kinds of taps to the shoe soles (pennies, bottle tops, metal plates and paper clips). Compare the prints.

More challenging for older children

Give them two or three tap shoes and several different-colored stamp pads.

Modifications for younger children

Use very small shoes, which are easier for younger children to grasp, and make sure the stamp pad has **washable** ink!

Polishing Shoes

Activity

▲ Place the shoes in a large basket.
▲ Place a small amount of shoe polish paste in the small container.
▲ Place the paste, the shoe polish brush and the rag on a tray beside the basket of shoes.
▲ Encourage the children to wear a smock or an old shirt to protect their clothes.
▲ The children use the polish applicator to put polish on the shoe.
▲ After the polish dries, they buff the shoe with the rag.

Questions you might ask

▲ How did you change the look of the shoe?
▲ Which shoes need polish and which do not? How do you know?

Extension

Place boots out for polishing.

More challenging for older children

Put out different colored shoes and different colored shoe-polish paste.

Modifications for younger children

This activity works best for younger children when it is first demonstrated by the teacher and then supervised carefully.

Interest area
Art

Science & math principle
Teaches about change as the shoes change when polished.

Science & math skills
Develops observation skills.

Science & math attitude
Teaches children respect for tools.

You need

2 or 3 old leather shoes
basket
nontoxic shoe polish paste
small container (like small jar lid)
shoe polish applicator brush
shoe polish buffing rag (old wash cloth)
tray
smock or old shirt

The Shoe Shuffle

Interest area
Group time

Science & math principle

Teaches children about patterns as they line up the shoes.

Science & math skill

Encourages thinking about different ways to organize shoes.

Science & math attitude

Develops cooperation as children work together.

▼

You need

shoes
ruler

Questions you might ask

▲ Why don't we have any shoes that measure 19 inches (48 cm) long?
▲ How would it feel to wear a shoe too big? too little?

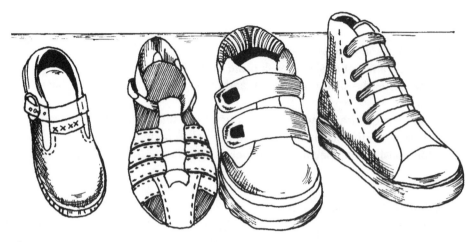

Activity

▲ The children take off one shoe and line up the shoes side-by-side.
▲ After all the shoes are in a line, the children put them in order by size, from smallest to largest (this will be noisy!).
▲ After they have all agreed on the order, measure the length of the smallest shoe with a ruler.
▲ Ask the children to find all the shoes that are that long.
▲ Measure the next shoe and have the children find all the shoes that are the same length. Continue until all of the shoes have been measured and grouped by size.
▲ Count how many shoes there are in each group. What is the smallest size with the fewest shoes? What is its length in inches (centimeters)? What is the largest size with the fewest shoes? What is the most common shoe length in the class?

Extension

Place a basket of shoes and a ruler in the Science and Math Center. The children can explore measuring shoes independently.

More challenging for older children

Ask the children to do this activity using **both** the width and the length of the shoes to put them in order.

Modifications for younger children

Do this activity in a small group over two days, since the younger children will not be able to wait for all of the children in the larger group to arrange the shoes by size. Ask the children to come back with the same shoes, if they can; then, measure and group the shoes on the second day.

How Many Socks Fill a Boot?

Activity

▲ Gather the children in a group around the boot and basket of socks.
▲ Ask them to predict how many socks it will take to fill the boot.
▲ List the children's names and their predictions on a chart.
▲ Stuff the boot with socks, one at a time, until it is full.
▲ After a child puts a sock in the boot, make sure it gets pressed down before another one is added.
▲ When the boot is full, take out the socks and count them to check their predictions.

Questions you might ask

▲ With what else could you fill the boot? How many would it take?
▲ What would happen if we tried to wear the boot with that many socks on our feet?

Extension

Place the boot and socks in the Science and Math Center for exploration. Add other large and small items that might be used to fill the boot so children can continue to make predictions and test them.

More challenging for older children

Do the same activity with a shoe and a boot. Compare and discuss the difference in the number of socks it takes to fill each.

Modifications for younger children

Use a small boot or shoe to reduce the time needed to fill either one. The children will not have to count so much to get the results.

Interest area
Group time

Science & math principle
Teaches children about the properties of boots.

Science & math skill
Develops comparison skills.

Science & math attitude
Develops children's honesty as they compare guesses with results.

You need

1 boot
large basket of old socks
chart tablet and marker

Interest area
Group time

Science & math principle
Teaches children about patterns as they observe similar characteristics of shoes.

Science & math skill
Encourages communication skills.

Science & math attitude
Develops children's cooperation.

▼

You need

children, with their shoes on
king-size white sheet
washable colored marker

Shoe Graph

Activity

▲ Mark an 8' x 8' (2½ m x 2½ m) square on the bed sheet with the marker. Section the square into a grid of 32 small 2' x 1' (60 cm x 30 cm) rectangles with the colored marker (see the illustration). Each vertical column on the grid will consist of eight 1' (30 cm) sections; each row will consist of four 2' (60 cm) boxes in which the children will stand according to their shoe color.

▲ Invite the children to stand in the appropriate column by shoe color. In other words, all of the children with blue shoes will be standing in one column, children with red shoes in another, etc.

▲ With this activity, the children are actually making a whole-body graph of the color of their shoes.

▲ Discuss with the children the results of the graphing: which shoe color is the most popular in the class?

▼

Questions you might ask

▲ How many colors (sizes) of shoes did we have in class?
▲ Is there another way to graph shoes?

Extension

Graph the shoes according to the following classification: boots, shoes that tie with shoelaces, Velcro closures, buckle closures and slip-on shoes.

More challenging for older children

After the children have graphed their shoes by color, have them graph by size. This will be noisy and confusing because many of the children will have to look at other children's shoes by standing beside them and comparing the lengths.

Modifications for younger children

Use two pairs of the children's shoes and place them on the grid in the appropriate column to demonstrate the activity. Give the children back their shoes and show them how to stand on the graph with their shoes on. Be sure to place the sheet with the grid in an activity center for them to practice graphing themselves.

Shoe Box Sorting

Interest area
Manipulatives

Science & math principle
Teaches children about patterns as they find similar characteristics of shoes.

Science & math skill
Helps children organize information about sorting.

Science & math attitude
Develops children's confidence.

▼

You need

shoes of various types (ballet slippers, rubber boots, cowboy boots, tennis shoes, snow boots, wing tips, loafers, dress shoes, sandals, golf shoes)
2 large cardboard boxes
paper
markers
tape or stapler

Questions you might ask

▲ Which type of shoes do you like the best?
▲ What would you use to make a shoe?

Activity

▲ Make two signs: "Work" and "Play"
▲ Attach the signs to the two boxes.
▲ The children sort the shoes by those that are best for work and those that are best for play. This activity is an introduction to grouping.

Extension

Group the shoes in other ways such as by size, type of material, color, appropriateness for cold or warm weather, etc.

More challenging for older children

Place two yarn circles, each approximately 3' (1 m) in diameter, on the floor. Overlap them so you have a Venn diagram. Have the children place work shoes in one circle, play shoes in the other circle, and shoes that could be worn for work and play in the space where the circles overlap.

Modifications for younger children

Before this activity is introduced, give children several days to explore the shoes by trying them on. After they become familiar with the shoes, do the activity using three or four pairs. Do it the following day with five or six pairs.

Heel Seriation

Interest area
Manipulatives

Science & math principles
Teaches children about patterns as they observe designs in heels.

Science & math skill
Develops organization skills.

Science & math attitude
Develops a desire for knowledge (curiosity).

▼

Activity

▲ The children arrange the heels on the tray from the largest to the smallest or smallest to the largest.

Questions you might ask

On what kind of shoe do you think this heel belongs? (Hold up one heel.) Why do you think the heels come in different sizes?

Extension

Seriate the soles of shoes (they are also found at the shoe repair shop).

More challenging for older children

Use blindfolds so the children seriate the heels by touch.

Modifications for younger children

Trace the shoe heels on two boards to make heel silhouettes. On one board, order the silhouettes from largest to smallest and on the other board, from smallest to largest. The children use the silhouettes to help them seriate the heels. When they can do this easily, put away the silhouette boards and see if they can seriate the heels without help. If they cannot, use three rather than six heels. Add one heel at a time as they master the activity.

You need

6 old shoe heels, different sizes (a shoe repair shop might donate them)
tray

Measuring Shoes

Interest area
Science and math

Science & math principle
Teaches about the properties of shoes.

Science & math skill
Encourages children to compare shoe sizes.

Science & math attitude
Develops children's cooperation.

▼

You need

a shoe-size measuring device (a shoe store may lend you one)
paper
pencil
chart tablet
marker

Activity

▲ The children use the measuring device to measure and record each other's foot size.
▲ Record all the children's shoe sizes as they measure each other and call out the numbers and widths.
▲ Create a chart with the list of the children's sizes.

Questions you might ask

▲ How many children wear the same size shoe?
▲ What would people look like if everyone wore the same size shoe?

Extension

Measure other children's and adult's feet. Which size is the most common?

More challenging for older children

Ask the children to discuss with their parents why they think we measure foot length in numbers (that do not represent inches or centimeters) and foot width in alphabetical letters.

Modifications for younger children

Do this in small groups or individually. Younger children are not able to wait for all of the children in the class to measure.

A Shoe "Home" for Critters

Activity

▲ Read *The Old Boot* by Chris Baines. This book is an excellent way to introduce the activity to the class. It is about a boot that was tossed out and became a home to many different insects.

▲ Place the shoe or boot outside in an area that is a little isolated, but where children can make frequent observations.

▲ Define the "no-walking zone" around the boot with a sign or low, string fence.

State the Problem or Question

What outside critters will make a home in, or under, our old boot (or shoe)?

State the Hypothesis

We believe a _____, a _____, a _____, and a _____ will come and live in or under our old boot. (Let the children predict what might come and make a list of their predictions.)

Method of Research

We will place the boot outside and we will not pick up the boot for four weeks. Then we will pick it up and wiggle it, and look beneath it to see what is living there. We will record our observations by listing, then counting, the critters we find. We will gently put back the boot after our observations. We will check the boot all year long to see the changes.

Checking the Hypothesis

We will look at our list of predictions and see if we were correct.

Results

We found a _____, a _____, a _____ and a _____ making a home in, or under, our boot.

Questions you might ask

▲ What might come to live in our shoe if we hung it in a tree?

▲ How is a critter's home different from ours?

Extension

Keep track of the changes in the boot itself. Make a list of what happens to the texture and color of the leather and other changes the children notice. This will help them focus on what nature does to things over time.

More challenging for older children

Encourage the children to draw a picture of what it is like under the boot.

Modifications for younger children

Place a hula hoop around the boot before you lift it. Have the children lay on their stomachs around the hula hoop to watch. Be sensitive to children who are frightened by little critters.

Interest area
Science and math

Science & math principle
Teaches children about diversity as they observe different insects.

Science & math skill
Develops observation skills.

Science & math attitudes
Encourages patience as children wait for insects to move into the shoe.

You need

The Old Boot by Chris Baines
1 dispensable large shoe or boot (to be a home for "critters")
string
sticks
chart tablet and marker

Interest area
Group time

Science & math principle
Teaches about the properties of different parts of a tree.

Science & math skill
Teaches children to compare by estimating.

Science & math attitude
Teaches children to respect nature.

You need

small shoe box
small items that will fit into the box:
- ✓ small twigs
- ✓ tree leaves
- ✓ pieces of tree bark
- ✓ several small tree roots

sign that says "What is in the Box?"

Predicting

Activity

▲ Each day, put one of the items listed in the shoe box.
▲ The children pass around the box, feel its weight and shake it.
▲ They guess what is in the box.
▲ After everyone has guessed, the children look inside and guess where the object came from (which part of the tree) and what it is made of.
▲ Then they guess how a tree might use it.
▲ Continue to do this until all the items have been examined.
▲ After the second or third day, the children will see the pattern and guess that the objects in the box come from trees.

Questions you might ask

▲ Have you ever seen anything like this before? (Hold up one item.)
▲ Which part of the tree do you like the best?

Extension

Visit a tree and guess its age. Ask the children to study the tree and estimate the number of branches, the number of leaves, how big around it is, the number of twigs on the branches and how tall it is.

More challenging for older children

Ask the children to think about five ways they could use the item in the box. Have them draw pictures of each way.

Modifications for younger children

Take them outside to locate where the twigs, leaves and other objects were found.

Twig Seriation

Activity

Note: When collecting twigs for this activity, start with a twig 4″ (10 cm) long and add twigs that are 1″ (2.5 cm) longer than the one before it.

▲ Place the twigs in the basket.
▲ The children use the tray to put the twigs in order by length, shortest to the longest and longest to the shortest.

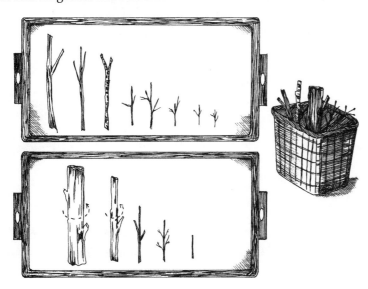

Questions you might ask

▲ How are the twigs alike? different?
▲ Can you think of a new way to put the twigs in order?

Extension

Ask the children to find other tree items they can put in order by size.

More challenging for older children

Ask the children to put the twigs in order of the size of their diameters from the thickest to the thinnest and thinnest to the thickest.

Modifications for younger children

Create a silhouette board for the twig seriation. Place the twigs, seriated from largest to smallest, on a copying machine and make a copy. Repeat with the twigs in reverse order. Place the copies (silhouette boards) on a tray. Have the children match the twigs to their silhouettes. Later, remove the silhouette boards so the children can seriate the twigs without the aid of the silhouettes.

Interest area
Manipulatives

Science & math principle
Teaches children to find patterns in the twigs.

Science & math skill
Teaches children to organize their knowledge of twigs by seriating them.

Science & math attitude
Develops a desire for knowledge (curiosity).

▼

You need

5 or 6 twigs of progressively longer lengths
large tray
large basket

Tree Objects Sorting

Science & math principle

Teaches about the interdependence between people and trees.

Science & math skill

Teaches children to organize knowledge by sorting.

Science & math attitude

Encourages children to respect nature.

▼

From a Tree   Not From a Tree

You need

assortment of items from a tree: acorns, nuts, twigs, leaves, roots, bark pieces, seeds, tree rings (a cross-section of a tree or branch), paper, wood scraps, sawdust in resealable plastic bag, doll furniture, jar of maple syrup, mulch in resealable plastic bag.

assortment of items that are not from a tree: plastic blocks, sponges, buttons, balls, pieces of fabric, bells, markers, playdough

large basket to hold all the materials listed above

large sorting tray

picture of a tree

picture of a tree with a line through it

marker

Activity

▲ Place the two pictures side by side on the bottom of the sorting tray. Each picture should cover half the tray.

▲ At the bottom of the tree picture, write "From a Tree" and at the bottom of the crossed-out tree picture, write "Not From a Tree."

▲ Put all the items listed above in the basket and place the basket beside the sorting tray.

▲ The children take turns picking items from the basket. They examine the item and decide whether it came from a tree or not. Then they place it on the appropriate picture in the tray.

Questions you might ask

▲ What would it be like if we did not have trees?

▲ Pretend you are a tree. What fruit would you grow on your branches? How would your fruit be used?

Extension

Make paper. Follow the instructions in *Good Earth Art* by MaryAnn Kohl and Cindy Gainer (page 192).

More challenging for older children

Add a third category that is neither part of a tree nor a product from a tree. For example, you might add a bag of dirt or grass, a glass or plastic jar, a bag of coffee, a clock, a paper clip, chalk and a piece of fabric. The third category is "Everything Else."

Modifications for younger children

Limit the items to be sorted to three or four in the two categories. Have the children do the activity in small groups so you can work with them as they sort.

Tree Listening

Activity

▲ Ask the children what they think they will hear when they listen to a tree with a stethoscope.

▲ Next ask them what they think they will hear when they listen just with their ears.

▲ After you have generated the lists, go outside with the children.

▲ They listen with the stethoscope and then with their ears.

▲ Check the results of what they actually heard against their predictions.

▲ Make a summary statement about their research (for example, we heard _____ sounds with the stethoscope and _____ sounds with our ears.)

Note: Some young children are not yet desensitized to the sound of their own heartbeats and may think they are hearing a heartbeat from the tree.

Interest area
Outside

Science & math principle
Teaches about the properties of trees.

Science & math skill
Develops observation skills.

Science & math attitude
Encourages a desire for knowledge (curiosity).

▼

You need

stethoscope (2 if possible)
tree
chart and marker

Questions you might ask

▲ If the tree does not have a heartbeat, how do we know it is alive?

▲ Do you think other trees might have a heartbeat?

Extension

Use the stethoscopes on other trees, bushes and plants.

More challenging for older children

This is a good activity to discuss how negative information is helpful in research. Information like: we know the tree does not breathe like we do because we cannot hear it, we know the tree does not have a heart like ours because we cannot hear it and we can know what something is by what it does not do.

Modifications for younger children

You may need several stethoscopes to maintain the interest of younger children in this activity. Have the children listen to themselves first. Also, have them examine and explore how the stethoscope functions before going out to listen to a tree. Do this activity without the prediction chart.

Mapping

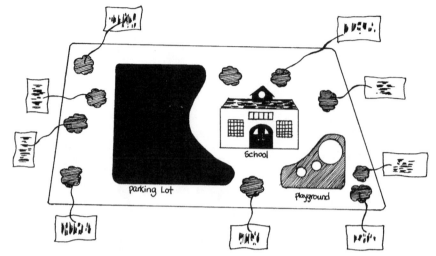

Activity

▲ Take a tour of the area surrounding the school and observe the location of the trees (go to a park if your school yard doesn't have enough trees).
▲ With the children's help, draw a map of the trees and number each one.
▲ Take a second trip on another day and get bark rubbings from the trees.
▲ As you go from tree to tree with the children, write the number of the tree on the map that corresponds to the rubbing taken from the tree.
▲ When you return from your tour, display the map and the rubbings.
▲ Tack yarn on the display to connect the tree drawing on the map and its rubbing. (If you have access to a Polaroid camera, take pictures of each tree.)
▲ As a class, make a list of the differences and similarities between the rubbings.

Extension

Do rubbings from different parts of the tree and compare them.

More challenging for older children

Group small pieces of bark from the trees in a Venn diagram. For example, one circle is for bark that is mostly light brown, one circle is for bark that is mostly dark brown and the overlapping area of the two circles is for bark that has both colors.

Modifications for younger children

Do this activity with two or three trees only. Take photographs of the trees, if possible, and add them to the map. The children may need help doing the rubbing.

Composting

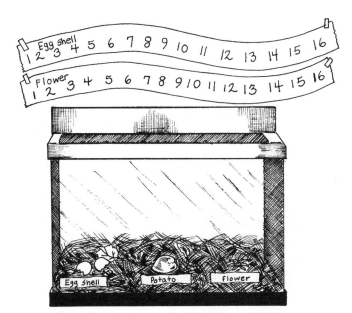

Interest area
Science & math

Science & math principle
Teaches about change as children observe the rotting process.

Science & math skill
Develops observation skills.

Science & math attitude
Helps children develop patience.

Activity

▲ Place twigs in the bottom of the aquarium. This allows air (oxygen) to circulate around the contents, encouraging the decay process.

▲ Dump in leaves, grass and other natural materials.

▲ Throw in leftover bread pieces and any vegetable products you have available. Do not use animal proteins or fats since they do not decay rapidly and might attract bugs or other unfriendly critters.

▲ Place several recognizable objects like an eggshell (thoroughly rinsed with water to remove the proteins), half of a potato, a large flower head, a twig and a mushroom in the front of your aquarium, right up against the glass.

▲ Label each object by taping its name above it on the outside of the aquarium. Use strips of adding-machine tape to make number lines for each object. Label the adding-machine tape with the name of the object.

▲ Attach the tapes to the wall above the aquarium.

▲ The children observe the recognizable objects throughout the school year and note on the adding-machine tape for each object the day that object began to decay.

▲ After you have established your compost aquarium, mist it well.

▲ Add a dozen earthworms. Subsequently, mist the compost material well every two or three days and stir it gently to avoid having worm "casualties" and disturbing the eggshell, potato, twig, flower and mushroom that are being observed.

▲ As you go through the school year, have the children add to the material in the compost. Let them mist and stir it.

▲ Ask the children to predict the number of days it will take for the objects at the front of the aquarium to decay. Keep track of each item that you are watching on the adding-machine number line.

You need

aquarium
twigs
grass clippings
leaves
flowers
bread pieces
eggshell
vegetable and fruit leftovers
water in a misting bottle
earthworms
labels
adding-machine tape
chart tablet
marker
tape

Questions you might ask

▲ In what way has the material in the aquarium changed?
▲ Why do you think the change is happening?

Extension

Add a large thermometer in the center of the compost to keep track of the temperature changes in the aquarium as the compost material decays.

More challenging for older children

Encourage the children to keep a descriptive journal of the changes that are occurring in the aquarium. This can be done individually or in small research groups.

Modifications for younger children

Trace the outline of each item that you are observing with a red marker on the aquarium glass (with the red-marker line as a reference shape, the children can easily notice changes in the object as it decays). Better yet, if you have a photo or a drawing of each object, place them on the front of the aquarium for comparison. Place a screen on the top of the aquarium to keep small hands out.

Leaf Graphing

Activity

▲ Take the children on a walk to gather leaves.
▲ Bring a basket for the children to fill with leaves.
▲ When you return to the classroom, the children graph the leaves on the graphing grid.
▲ They decide which character- istics they want to use to graph the leaves (patterns within the leaves, shape, texture or color). For exam- ple, you might have leaves that are rough, leaves that are long and slender, leaves that have several segments, leaves that are feathery and leaves that are very small. (You can add the leftover leaves to the compost aquarium.)

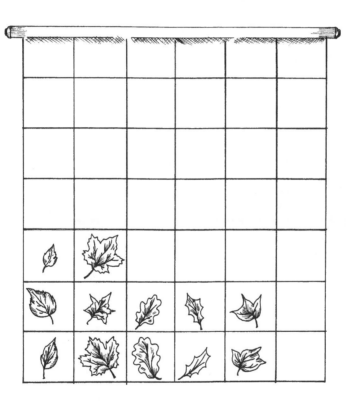

Interest area
Science and math

Science & math principle
Teaches children to recog- nize patterns in leaves.

Science & math skill
Teaches children to commu- nicate their findings by graphing.

Science & math attitude
Encourages children to be comfortable with science and math.

You need

basket
window-shade graphing grid from Rock Graphing, page 197
variety of leaves

Questions you might ask

▲ What is alike about the leaves? different?
▲ How do you know that this is a leaf?

Extension

After graphing the leaves as a whole class, give the chil- dren graph paper to graph individually all of the leaves they collected.

More challenging for older children

Graph the leaves one way, then ask the children to look at them differently and graph them another way. For example, they might try graphing leaves accord- ing to where they come from: a bush, a tree or a small plant.

Modifications for younger children

Graph only a few leaves at one time. Add to the graph at a second sitting, and then again at a third.

Tree Measuring

Interest area
Science and math

Science & math principle
Teaches about the properties of trees.

Science & math skill
Teaches children to compare by measuring the circumference.

Science & math attitude
Teaches children to respect nature.

▼

You need

string
scissors
trees
tape measure
chart tablet and marker
large sheet of paper

Activity

▲ Take the children outside and find a favorite tree.
▲ Ask them to guess how big around it is (use the word "circumference").
▲ Record their guesses.
▲ Using string, the children measure around the tree.
▲ Place the string beside a retractable tape measure and find the circumference in feet and inches (meters and centimeters).
▲ Compare the result to the guesses.
▲ Take the string inside and lay it in a circle on the large sheet of paper. The children trace around the string.
▲ Cut out the circle and talk about the circumference of the tree. Mark the center of the circle. Have the children draw a line through the center to the opposing edges of the circle and introduce the word "diameter." Measure the diameter. Post the tree circumference and diameter tracing in the Science and Math Center for the children to write "tree" words on.

Questions you might ask

▲ Why do you think it is important to know how big around a tree is?
▲ How could you find out how big around different trees are?

Extension

Measure other trees and compare their different circumferences and diameters.

More challenging for older children

Measure the circumference of the tree at different heights and do the same activity with each measurement as above. Compare the circles and decide why they are graduated in size (e.g., why is the tree bigger around at the bottom than at any other point?).

Modifications for younger children

Let the children measure how many children will fit around the tree. Later, they measure with the string.

▼

Tree Shadows

Activity

▲ Eight times during the day, the children take a stake outside and hammer it into the ground at the top of the shadow of a tree.
Note: Be sure to carefully supervise the use of the hammer and stake.

State the Problem or Question
How will the shadow of a tree change during the day?

State the Hypothesis
The children will predict what they think will happen to the shadow of the tree during the day. They will list their predictions.

Method of Research
We will start first thing in the morning by placing a stake at the top of the tree shadow. We will write the time we started on the stake. We will mark the tree shadow with a stake the same way every hour, and we will write the time on the stake.

Checking the Hypothesis
We will look at our stakes and observe their different placements to see if our predictions were correct.

Results
We found that the shadow of the tree _____.

Interest area
Science and math

Science & math principle
Teaches how a tree shadow changes because of the sun's movement (cause and effect).

Science & math skill
Teaches children to compare their knowledge as they chart changes.

Science & math attitude
Encourages children to be curious.

You need

nearby tree
8 wooden stakes
hammer
permanent marker
chart tablet

Questions you might ask

▲ What do you think made the shadow move?
▲ Would you be able to see the shadow if it were cloudy?

Extension

Do this with two other trees that are different heights and see if the changes are the same.

More challenging for older children

Encourage the children to do individual record keeping of the changes in the tree shadow. They can measure the distance between the stakes as the shadow changes.

Modifications for younger children

Start first with the children examining their shadow when they move. Stake out the shadow movement. Talk about it and their observations.

Filming Clouds Across a Window

Interest area
Art

Science & math principle
Teaches about cause and effect as children observe moving clouds.

Science & math skill
Encourages children to communicate their information by making a picture.

Science & math attitude
Teaches children to be curious about clouds.

You need
window
clouds in the sky
leftover laminating film, approximately 8" x 11" (20 cm x 28 cm)
permanent markers in different colors
clear tape

Questions you might ask
▲ What do you think moved the clouds?
▲ Why do clouds move fast sometimes and slow at other times?

Activity
▲ Tape the clear laminating film to the inside of a classroom window.
▲ A child looks through the film to find one cloud to observe during the day.
▲ She then traces the outline of that cloud on the film with permanent marker.
▲ About a minute later, she makes her second observation. If she can still see the cloud through the laminating film, she traces its outline again using a different color of marker. If this outline overlaps the first one, she will see that her cloud moved very slowly and in which direction it moved. If the cloud is gone, she will see that it moved very fast.
▲ In some instances you may have five observations in five minutes and in others, only one. Talk about how the cloud moved.

Extension
Ask several children to do the activity on the same day. Compare the different shapes, results and movement of the cloud drawings.

More challenging for older children
Use a ruler to measure the number of inches (centimeters) the cloud moved across the laminating film. Have them speculate on how far they think the cloud actually moved across the sky.

Modifications for younger children
Before beginning this activity, go outside and have the children sit or lay down on the ground. Have them watch clouds move across the sky. Repeat this introduction process several times during the week, then do the activity.

Hear Gusty Chimes

Interest area
Outdoors

Activity

▲ Before the children arrive, cut pieces of strings into different lengths.

▲ Tie a washer to each end.

▲ Divide the class into small groups.

▲ Each group makes a wind chime by tying the strings with attached washers to the strawberry basket. For a little variety, you might want to add small bells, but this is optional.

▲ Many of the children may be unable to tie with the string. If this is the case, work with each group individually at different times during the day.

▲ When everyone is finished, tie a length of string to each basket and hang them near the classroom window or in a nearby tree so the children can hear the wind make them chime. If that is not possible, hang them in the classroom and turn on an oscillating fan.

Hang the basket on a branch of a tree and listen as the wind makes music.

Science & math principle

Teaches about energy as children observe mild and strong winds.

Science & math skill

Encourages observation skills.

Science & math attitude

Develops children's curiosity.

▼

You need

4 to 6 strawberry baskets (1 basket for every 4 to 5 children)
string
washers of different sizes
small bells (optional)

Extension

Examine a variety of commercially made wind chimes. Compare sounds.

More challenging for older children

Have each child make her own chime. Allow her to choose the washers, tie the string and hang the string on her strawberry basket.

Modifications for younger children

Make one basket for the whole class. Most younger children cannot tie. They have not developed the fine motor skills to accomplish this kind of task. The benefit to them is derived from watching and listening to the wind chime.

Questions you might ask

▲ When there is no wind, do you hear the chimes? Why?

▲ How do the chimes make music?

Tossed Leaves

You need

basket of leaves
couple of rakes
retractable tape measure,
   25′ (7 meters) long
large ball of string
scissors
masking tape

Questions you might ask

▲ Why do you think the leaves traveled different distances?
▲ Why do you think some leaves flipped and tumbled as they floated to the ground?

Activity

Note: Do this activity on a windy day.
▲ Find a nearby tree that sheds its leaves and ask the children to rake them into a pile. If you do not have such a tree, bring a basket of leaves from home and dump them in the school yard.
▲ One child tosses the leaves into the air from one of several different points in the school yard (near the swing, near the slide, behind the sand box). Watch them float to the ground.
▲ Use the ball of string to measure the distance from the first pile to a leaf farthest from the pile. Cut the string at that point. Take a piece of masking tape and write "first toss from the swing." Attach the masking tape to the string.
▲ Move the pile to a second point and a different child tosses an armful of leaves. Watch them float to the ground. Again, measure with the string, cut and mark it as with the first toss but write "second toss from the slide."
▲ Repeat this exercise several times (if you run out of leaves, have the children rake up the leaves again for the next toss).
▲ Talk about wind currents carrying the leaves and the strength of the wind.
▲ When you have at least four string lengths from four tosses, go back to the classroom and compare the length of the strings.
▲ Measure each string with the tape measure. Talk about why there is a difference.

Extension

Try this activity where obstacles will make an obvious difference in the distance and the dispersion of the leaves (for example, toss them from behind a building toward a passageway between two buildings or where the wind swirls). If you choose a day when the wind is strong, remind them to keep their mouths closed as they watch the results!

More challenging for older children

Extend the tape measure to its full length. Show it to the children. Then let the children do this activity individually, with a handful of leaves. Have each child choose a spot on the school yard that is at least twice as long as the extended tape measure from another child. Have them estimate the distance, then make their tosses. Have the children compare the string lengths from their individual tosses and discuss why they are different and how the wind is affected by objects that block the wind.

Modifications for younger children

Let the children toss leaves in the air many times and then rake them up. Let them have these experiences before you use the string to measure. Have them do this activity only twice, so they have only two strings to compare.

Actively Windy

Activity

Note: Do this activity on a very windy day.

▲ The purpose of the following five "mini-activities" is to expose the children to the forces of the wind and to help them draw conclusions about how the wind affects our lives (sometimes it makes our lives easier, while at other times it makes our lives more difficult).

✓ On a windy day, sit in a spot outside where they can see trees, shrubs and bushes, people walking, clouds, and tall grass, as examples. Move around to find things to observe. Ask them to notice what happens to people and to objects when the wind blows.

✓ Two children stand facing each other about 10 feet (3 meters) apart, one with his back to the wind, the other facing the wind. Give one child a ball and have the two toss the ball back and forth. Talk about how far the ball goes with the wind at your back and how much more difficult it is to toss it into the wind.

✓ Take the tape and tape recorder outside on a blustery day. Tape the unusual sounds the wind makes, such as whistling, rustling leaves and moving playground equipment.

✓ Take a sturdy umbrella outside on a not-too-windy day. Show the children how to point the umbrella into the wind to keep the wind from rushing up under the canopy and collapsing the umbrella. Ask them to feel the force of the wind. Could the wind knock them down? What would it be like to be a sailboat on the sea?

✓ Have two children face each other about 10 feet (3 meters) apart. Have them try to talk to each other. Who is the easiest to hear, the child with the wind at her back or the one with the wind in her face?

Extension

Do the same things on a day when there is little or no wind.

More challenging for older children

Encourage the children to make a wind investigation book recording all their observations about the wind. They can add drawings about the effects of wind on the things they have seen.

Modifications for younger children

Do this activity on different days. Be very careful when they handle the umbrella. Hold on to it with each child to feel the force of the wind. If the ribs of the canopy protrude, apply corks.

Interest area
Outdoors

Science & math principle
Teaches about the energy of the wind.

Science & math skill
Encourages observation skills.

Science & math attitude
Develops children's curiosity.

You need

ball
umbrella
blank audio tape
battery-operated tape recorder

Questions you might ask

▲ How does the wind help us?
▲ What would you feel like if you were the wind?

Windy Muscles

Interest area
Science and math

Science & math principle
Teaches that wind can move some objects and not others (cause and effect).

Science & math skill
Develops observation skills.

Science & math attitude
Encourages children to be comfortable with science and math.

▼

You need

large rock
block
book
doll
basketball
baseball
piece of paper
large piece of fabric
piece of newspaper
electric fan with high and
    low speeds (with a plastic
    fan guard securely in
    place)
large tray
chart tablet and marker

Questions you might ask

▲ What was similar about the objects that were moved? different?

▲ Would the baseball have moved if it was square? Why did it move?

Activity

▲ Before the children arrive, list each object on a chart tablet. Beside each one, write the words "yes" and "no."

▲ To begin the activity, ask the children which object on the chart tablet could be moved by the wind blown by the fan. Review each object.

▲ Reach a decision as a group. Circle the group's choice.

▲ Place all the objects on a tray.

▲ Test each object by placing it in front of the fan, first at the low speed, then at the high speed. Talk about the differences in forces of the wind at low and high speed and how the objects were affected (for example, how far they were moved by the force of the wind). Talk about the wind from the fan and the wind outside. Are "inside" wind from a fan and "outside" wind made by nature very different?

Extension

Encourage the children to gather a variety of other objects. Have them make their predictions, then test them.

More challenging for older children

Ask the children to guess and test each item first with a hand-held fan (wave back and forth) and then with the electric fan. Discuss the results of what the hand-held fan will move and what the electric fan will move.

Modifications for younger children

Let the children hold each object beforehand. This will help them grasp the concept of weight (it may be difficult for them to decide just by looking at the objects). Start by testing four or five objects initially; then add other objects to test as the children comprehend the force-weight relationships being demonstrated.

The Big Straw Blow

Activity

▲ Before doing this activity, make a small chart for the children to use when predicting and testing. Duplicate it for all of the children in the class.

▲ Place the container of small objects, the straws and copies of the chart on the tray.

▲ The children use a straw to blow small items across a tray, table or any flat surface.

▲ First they predict which objects they think that they can blow across the tray and then circle their predictions on the chart.

▲ Then they test their predictions by blowing on the items with a straw. On the chart, next to the picture of the item, they put a checkmark for "yes" if they were able to move it by blowing, and an X for "no" if they could not.

▲ After you have demonstrated this activity, place it in the Science and Math Center.

Questions you might ask

What could you move the easiest? Which was the hardest to move? In what way were the objects you could move easily alike?

Extension

Encourage the children to collect other items to test. Allow them to draw their own pictures on the chart.

More challenging for older children

Ask the children to make a list of the characteristics of the objects that moved easily and the characteristics of the objects that did not move.

Modifications for younger children

Cut the straws in half. This makes the distance for the air to travel shorter, making more wind force available to make the objects move. Have three or four objects for them to test. Do not use the prediction and testing sheet.

Interest area
Science and math

Science & math principle
Teaches about the energy of the wind.

Science & math skill
Teaches children that observation is a tool used to gather information about things.

Science & math attitude
Encourages children to be honest.

▼

You need

straws (at least 1 per child)
small objects like an acorn, penny, cotton ball, paper clip, small rock, game piece, scrap of paper, pencil, leaf
tray
container
crayons
chart tablet and marker

Which Way Is the Wind?

Interest area
Science and math

Science & math principle
Teaches about the energy of the wind.

Science & math skill
Develops observation skills.

Science & math attitude
Encourages children to be comfortable with science and math.

▼

You need
12″ (30 cm) piece of thick yarn (ponytail yarn has a much wider gauge than other yarns)
paper
markers
copy machine
clipboard

Questions you might ask
▲ Why do you think wind direction is important to know?
▲ How many days did we have when the wind changed direction?

Activity
Note: This activity works best in the fall or early spring.
▲ You may need help from your maintenance staff to find a high place outside to tie the yarn. It needs to be visible from the classroom and in a place where the wind can move it.
▲ Draw a picture of the object to which the yarn has been tied and make several copies. Put one copy on a clipboard.
▲ The children watch the yarn to see which way the wind is blowing.
▲ Each day, ask one child to draw the yarn (on one of the duplicated drawings you made) exactly as he sees it.
▲ Date the drawing and post it in the classroom.
▲ Repeat this activity, skipping days randomly, but watch for changes in the weather. Post each subsequent day's drawing and date it. Place it next to the last observation drawing.
▲ Ask the children to look at the drawings every day and to discuss what patterns they see. Talk about wind direction in general terms without referring to the drawings. Ask the children how they would tell someone over the phone which way the wind was blowing (without using directional words like north and south).

Extension
Keep track of the wind direction by checking it once a week and looking at the pattern over the entire school year. What patterns emerge?

More challenging for older children
Encourage individual children to keep track of the changes by drawing a picture each day of what was seen.

Modifications for younger children
Delete the record-keeping component of this activity. Each day have them talk about where the yarn is pointing and what is making it move.

▼

Windy Photographs

Activity

▲ Before doing this activity with the children, glue all of the photographs or magazine pictures to construction paper.
▲ Cut out the pictures, leaving a border of construction paper around each.
▲ Laminate or cover the pictures with clear contact paper.
▲ Place them in the basket.
▲ Make the signs, "Safe Winds" and "Dangerous Winds."
▲ Place the basket on a tray with the signs and place the tray in the Science and Math Center.
▲ The children sort the "safe wind" photographs from "dangerous wind" photographs by placing them under the signs.

Questions you might ask

▲ What makes strong wind dangerous?
▲ What would you do if a strong wind started to blow on you?

Extension

Investigate the beneficial and the destructive effects of wind on the earth.

More challenging for older children

Delete the photographs. Ask the children to imagine strong wind scenes that fit into the two categories and then draw pictures for the rest of the class to sort.

Modifications for younger children

Before you do this activity, let the children feel the force of the wind from a fan on high and low speeds or go outside on a windy day and encourage them to look at moving tree branches, waving flags and floating leaves.

Interest area
Science and math

Science & math principle
Teaches about the energy of the wind.

Science & math skill
Teaches children to organize information by sorting.

Science & math attitude
Teaches respect for nature.

▼

You need

different photographs or magazine pictures of the wind in action (for example, the wind blowing a kite or clothes on a line, trees bent from the wind, tornadoes, hurricanes and driving rain storms)
construction paper
glue
scissors
clear contact paper or laminating machine
basket
markers
tray divided in half by colored tape
"Safe Wind" and "Dangerous Wind" signs

Interest area
Science and math

Science & math principle

Teaches about the interaction between the wind and the clothes.

Science & math skill

Develops observation skills.

Science & math attitude

Helps children learn to be patient.

▼

You need

lightweight, adult-size shirt (cotton is best)
heavy child-size pair of blue jeans
dishpan of water
outside clothesline or chain-link fence
6 clothespins
chart tablet and marker

Questions you might ask

▲ Which one dried first? Why?
▲ Do you think they would have dried faster in a clothes dryer? Why?

Air-Dried Duds

Activity

▲ On a windy day, have the children saturate the cotton shirt and the blue jeans with water.
▲ Ask them to predict which will dry first, the shirt or the jeans (children usually think that the size of an object determines how long it will take to dry; the bigger garment will dry fastest.)
▲ Write their predictions on a chart.
▲ Talk about how the wind moves air through the clothes to dry them. Ask where they think the water goes.
▲ Place the two items of wet clothing on a clothesline or a fence, securing them with clothespins.
▲ Check the clothes frequently during the day until one garment is dry.
▲ Ask the children to check to see if their predictions were correct.

Extension

Ask the children to make a list of all the wet objects and places they have noticed on a rainy day. They see lots of wet things on rainy days. Have them write estimates of how long it takes for these things to dry. How long does it take a sand box to dry? The playground? How long for a street to dry? What about a wet car? Why do objects dry faster or slower? What are the characteristics of an object that make it slow to dry?

More challenging for older children

Ask the children to describe why the shirt dried first and why the jeans took longer. Use a clock to record the time it took for each garment to dry.

Modifications for younger children

Compare the time it takes to dry a child-size, lightweight shirt and a heavier garment from a doll.

Balloony Movement

Activity

▲ Tie a balloon to the classroom ceiling, high enough that the children cannot reach it with their hands.
▲ Place the fans on a tray.
▲ Number the fans, starting with #1.
▲ On a large sheet of paper, write fan #1, fan #2, fan #3, and so forth.
▲ The children guess which fan will move the balloon in the quickest time when placed beneath it.
▲ They test their guess by waving the fan a couple of times and counting (one...and... two...and...three...and...four...) until the balloon moves. When the balloon moves, the child will stop counting.

Question or Statement of the Problem
Which fan will make the balloon move the quickest?

State the Hypothesis
We think fan # _____ will make the balloon move the quickest.

Method of Research
We will use each fan and wave it three times underneath the balloon. We will count how long it takes for the wind to be moved by the fan to the balloon. We will keep counting (one...and... two...and...three...and...four...) until we see the balloon move. When the balloon moves, we will stop counting. We will write the last number we counted on the sheet next to the number of the fan that was used. We will repeat this procedure for each of the fans on the tray.

Checking the Hypothesis
When we have tested all the fans we will look at the chart to see which fan has the lowest number beside it. The fan with the lowest number is the one that moved the balloon the quickest. We will see if that fan is the same one we predicted would move the balloon the quickest.

Results
We found that fan # _____ moved the balloon the quickest.

Extension

Do the activity with balloons of different sizes. Compare results.

More challenging for older children

Redo the experiment to find out which fan makes the balloon move for the longest period of time.

Modifications for younger children

Do not record predictions or results on the paper sheet. Have children guess which fan will make the balloon move the quickest and then talk about their discovery.

Interest area
Science and math

Science & math principle
Teaches about the energy of moving air.

Science & math skill
Teaches children to compare as they try different ways to move the balloon.

Science & math attitude
Encourages a desire for knowledge (curiosity).

You need

different sizes of hand-held fans
blown-up balloon, tied to a string
chart tablet and marker

Questions you might ask

▲ How did you make the balloon move?
▲ What would happen if you used a hair dryer? Would it be quicker or slower? Why?

Chapter 4

Assessing Science and Math

If children are exploring and manipulating materials, moving around the room, working in small groups and generally doing their own scientific investigation, how do we know if they are learning anything? That is a good question. Here is another: How do we justify this approach to parents and other teachers? This kind of classroom organization is far less structured than what they are used to. Processes such as investigating, estimating, testing guesses and not always getting "right" answers are emphasized, so how do we know what children are learning, and how do we document the progress they make?

These questions all relate to assessment (simply put, finding out what children know), a very important aspect of teaching. Why? Because assessment influences curriculum; that is, if assessment points out that something is not working, you can try something else right away! If you see a child struggling with a retractable measuring tape, pull out a folding carpenter's rule or a piece of string. Do it immediately. Why wait to test the child on measurement, then make the change? That may only breed the child's frustration and discouragement. Trying something that works reflects on-going assessment and evaluation.

I use two ways to document children's growth and development. First, I gather **work samples** that show children's skills in concept work or mastery. Second, I write **anecdotal records**. I use other tools for gathering information, but these two are at the heart of my assessment program. I gather information about the children **while they are working**. This method is called *portfolio assessment* because varied materials are gathered through the school year and often saved in a pocketed folder. With portfolio assessment,

you can get a more complete "portrait" of a child's development over time. No doubt you are asking at this point, "How do I assess science and math?" Another good question! There are several ways to collect information that the children are learning science and math principles, skills and attitudes. Let's look at them.

Assessment Using Work Samples

There are many activities in this book that lead to a piece of paper being produced by the child, such as a picture, chart or graph. The paper products show if a child has learned a particular skill or concept. For instance, if a child draws a sequence of pictures of the tree and its shadow as it moves across the ground, save the picture to show he was able to communicate what he had learned (which is quite a bit if you stop to think about it!). This sequence of pictures shows that he can **communicate** (an important science and math skill) his research or observation findings. He also has shown an **awareness of time** (a science and math principle), especially if the shadow sequence is in order. Place the drawings in his portfolio. It will demonstrate to you and to others that this child is learning important science and math principles, skills and attitudes.

Graphs or charts that note the child's participation can also be added to a portfolio. For example, in the apple graphing activity, copy the graph and place it in each child's portfolio. Write each child's name above the apple product she chose. In doing so you can see the child's involvement: she understands the **properties** (a science and math principle) of apples and she can **communicate** (a science and math skill) that understanding by graphing.

Any piece of paper that is created in the natural process of doing the activity can be used to expand the information base about children and their growth and development in science and math. I have asked children to make their own graphs and tally sheets, draw their own interpretations of events, write their own predictions and list their own descriptions of events. These writings are documentation of what they are learning. Add them to the children's portfolios. Look throughout the classroom for science and math learning. In the photograph below, you see Marsela painting a rainbow. She is showing she knows several things.

First, she is able to draw a **model** (a science and math principle) of a rainbow. She accepts the representation of a rainbow for the actual one. Second, she shows a grasp of **cause and effect** (a science and math principle) as she applies paint to a page. And third, she is able to **communicate** (a science and math skill) her understanding through painting. Carefully fold the painting (paint-side inward) in the portfolio, or take a color photograph of the painting and put the photograph in her portfolio. Not only is it an example of her painting, but it shows science and math learning as well.

Assessment with Audio Tape

Many things cannot be documented with work samples. In early childhood pencil and paper tasks are rare (and mostly inappropriate), so other methods of information gathering are necessary. Tape recording is one such method. Children can retell on audio tape the event sequence of cooking applesauce or of planting a tree. This captures the children's understanding of **change** (a science and math principle) and their ability to **communicate** (a science and math skill) this change. Use audio tape to record the children's descriptions of what they are observing, their feelings about their observations and their conclusions. Not only will you be getting science and math information, but you will be listening to how they think and construct sentences, and how well they **organize** (a science and math skill) their thoughts and express their understanding.

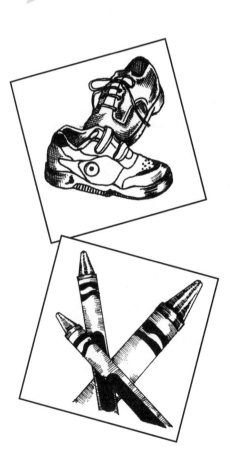

Sound complicated? Expensive? Well, audio tape recordings do not have to be either. Buy old tapes from garage sales or flea markets. Use clear scotch tape to tape over the two holes in the rear corners of the cassette and reuse them. Put a sticker on each tape with one child's name on it. Collect all the tapes in a basket and place it within easy reach of the children. Teach the children how to work a tape recorder. Teach them also not to rewind their tapes, but to start each time where they left off. Teaching them takes some time, because they love to hear themselves, but it is worth the effort you make. Before you start an assessment tape, let the children practice. Explain that they can listen to the whole tape when they are through. When they have finished and before they listen to the tape, pull off the scotch tape so they cannot record over their assessment recording. When you are not assessing the children, set up the audio recording area as an independent activity for the children.

Assessment with Photographs

Assessment with photographs is expensive, but worth it. I talked my own Parents and Teachers Association into buying the film and reimbursing me for the cost of film development. I used my own camera. The arrangement worked well.

Using photographs is another way to document the children's progress. There are so many events that cannot be gathered in any other way, as shown in the photograph below.

One example: if a child estimates the number of popcorn kernels in a jar and counts them on a number line to verify his estimate, a photograph can capture his understanding. It shows he understands **estimating** (a science and math skill) and that he knows how to **check his results** (part of the scientific method) using a number line.

When a child listens to the chimes made as she strikes silverware tied to headphones, she is showing an understanding of **cause and effect** (a science and math principle) and the science and math skill of **observing**. The following photograph captures this effectively.

When children are able to put the ears of corn in order by size, they are showing an understanding of the science and math principle of **diversity** (of a corn set) and an ability to **organize** (science and math skill) the ears by seriating them.

Any three-dimensional structure, like a cup construction, is difficult to save in a portfolio. Photograph it. It shows that children are able to **communicate** (science and math skill) their ideas and thoughts about how they "see" assembled structures. It also shows **respect** (a science and math attitude) for the equipment and materials being used. Photographs are just another tool to document children's growth and development across the curriculum.

Assessment through Anecdotal Records

Keeping anecdotal records is another way to gather information, and probably the one most frequently used in early childhood classes. These are brief, narrative statements of what a child is doing at the time. The narratives are factual statements of behavior, free of opinions about the behavior. They are non-judgmental observations made by the teacher of what she sees and hears while a child is performing a task in the natural setting for that task. There is no "setting up" to do for an anecdotal record. It describes what occurs naturally.

It is very difficult to get "evidence" that a child is curious, cooperative and patient. Many of these science and math attitudes can only be documented with anecdotal records. This kind of record keeping is just another tool used to document growth and development.

The following photograph will help you understand what was happening at the time I wrote my anecdotal record for this activity.

Marissa   10/12/__   Center Time,
Art Center   9:45 am

After introducing the corn-cob-rolling activity in group time, Marissa chose the Art Center. She rolled the cob in the paint, and she rolled it on the paper using her left hand. She said, "Look, look! It made corn bumps on the paper. It looks like the corn after we eat it."

When I reread Marissa's anecdotal record, I looked at what it told me about Marissa's learning. Marissa followed multi-step directions. She was aware of **cause and effect** (a science and math principle). She was able to **communicate** (a science and math skill) her understanding and she was **curious** (a science and math attitude) about the process in which she was engaged. She recalled the experience of eating corn, transferring that knowledge to this situation. Marissa was able to build on that knowledge.

Another example, shown in this photograph, is grinding corn in a Molcahette (a mortar-and-pestle-like tool).

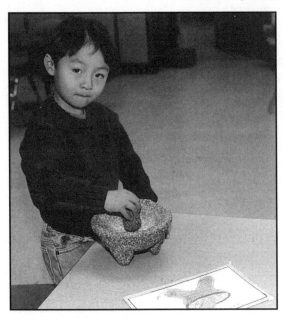

Li   10/20/__   Center Time,
Science and Math Center   9:20 am

Li was grinding the corn. He said, "Look, teacher! I made the corn look like flour. You told me! Now I told you...see the flour! It looks like the flour for the bread we made. Can we eat it?"

This anecdotal record told me that Li understood the science and math principle of **change**. He recognized that the corn will become flour (maza) when ground. He made **comparisons** (a science and math skill). He saw the similarities between the wheat we had ground earlier and the corn flour. He was able to remember an earlier experience and relate it to his new learning.

The examples are endless. Every interaction between a child and equipment or between one child and another develops skills and concepts. We need to know what to look for and how to record it in order to share the children's growth and development with others.

Assessment is such an important part of science and math that we need to set up an organized way (a protocol) to assess each activity. The use of work samples, anecdotal records, audio tapes and photographs is such a protocol. Procedures can be put in place through which we can capture children's growth and development over time.

Assessment Influences Curriculum

Remember that ongoing assessment drives curriculum. When something is not working, change it. Do not wait for the next assessment period to conclude that the child is not progressing. Use the information gathered to refocus your teaching. Simplify or increase the level of difficulty as you observe the children succeed or struggle to succeed. Share the information with others and use it to document the aspects of your program that work well.

Index

C

S